UNIVERSAL ACCESS
AND ITS ASYMMETRIES

Information Policy Series
Edited by Sandra Braman

The Information Policy Series publishes research on and analysis of significant problems in the field of information policy, including decisions and practices that enable or constrain information, communication, and culture irrespective of the legal siloes in which they have traditionally been located as well as state-law-society interactions. Defining information policy as all laws, regulations, and decision-making principles that affect any form of information creation, processing, flows, and use, the series includes attention to the formal decisions, decision-making processes, and entities of government; the formal and informal decisions, decision-making processes, and entities of private- and public-sector agents capable of constitutive effects on the nature of society; and the cultural habits and predispositions of governmentality that support and sustain government and governance. The parametric functions of information policy at the boundaries of social, informational, and technological systems are of global importance because they provide the context for all communications, interactions, and social processes.

A complete list of the books in the Information Policy series appears at the back of this book.

UNIVERSAL ACCESS AND ITS ASYMMETRIES

THE UNTOLD STORY OF THE LAST 200 YEARS

HARMEET SAWHNEY AND HAMID R. EKBIA

THE MIT PRESS   CAMBRIDGE, MASSACHUSETTS   LONDON, ENGLAND

The MIT Press would like to thank the anonymous peer reviewers who provided comments on drafts of this book. The generous work of academic experts is essential for establishing the authority and quality of our publications. We acknowledge with gratitude the contributions of these otherwise uncredited readers.

This book was set in Stone Serif by Westchester Publishing Services. Printed and bound in the United States of America.

Library of Congress Cataloging-in-Publication Data

Names: Sawhney, Harmeet Singh, 1960– author. | Ekbia, H. R. (Hamid Reza), 1955– author.
Title: Universal access and its asymmetries : the untold story of the last 200 years / Harmeet Sawhney and Hamid R. Ekbia.
Description: Cambridge, Massachusetts : The MIT Press, [2023] | Series: Information policy | Includes bibliographical references and index.
Identifiers: LCCN 2022006422 (print) | LCCN 2022006423 (ebook) | ISBN 9780262544559 (paperback) | ISBN 9780262372978 (epub) | ISBN 9780262372985 (pdf)
Subjects: LCSH: Digital divide—United States—History. | Poor—Information services—Government policy—United States—History. | Community information services—United States—History. | Discrimination in municipal services—United States—History.
Classification: LCC HM851 .S239 2023 (print) | LCC HM851 (ebook) | DDC 303.48/33—dc23/eng/20220425
LC record available at https://lccn.loc.gov/2022006422
LC ebook record available at https://lccn.loc.gov/2022006423

10 9 8 7 6 5 4 3 2 1

To Mee-Young
—Harmeet

To Mahin
—Hamid

CONTENTS

SERIES EDITOR INTRODUCTION

Sandra Braman

In the final work of his life, Edward Said wrote about "late style," address-ing the question of what it is that great artists do towards the end of their lives that is different from what they did before. That question is one that should be asked of scholars as well. *Universal Access and Its Asymmetries: The Untold Story of the Last 200 Years* is an unusual and highly original book by two strong and well-published senior scholars that takes up, per-haps without knowing it, Said's challenge by asking quite new questions about a subject on which much has already been written including, nota-bly, by the first author. In doing so, Harmeet Sawhney and Hamid Ekbia offer multiple contributions. In addition to significantly expanding the dimensions through which we should think about universal access in any domain, they provide insights into the nature of infrastructure, offer new methods for policy analysis, and provide insights into major sociological transformations.

The book examines the histories of seven systems for which universal access has become a policy goal in the United States over the last two hundred years—the postal service, education, libraries, electricity, tele-phony, broadcasting, and the internet. The authors start from the simple but stunning observation that analysis of this issue up to now, with its familiar focus on difficulties for systems and advantages for individuals, leaves out half the story: there are also gains for systems and travails for individuals and communities when service is extended. Filling in these missing quadrants turns network analysis on its head, for it requires looking not only at relations generated by links, but also at what happens, often very intimately, within the nodes. Both through explicit comment and by model, Sawhney and Ekbia also provide a critique of what are best understood as research industries, the transformation of social issues into jobs and institutional programs that become self-perpetuating in their inertia, regardless of effective productivity or utility.

The authors are unusually well read—at last, H. L. Mencken and Joan Didion come to information policy! Their use of historical materials from one and two centuries ago, whether news items, scholarly writing, or government documents, are valuable not only as support for their arguments, but also, often, as revelatory treats. Contemporary community-based voluntary efforts to build mesh networks to extend Internet access have predecessors in early-twentieth-century telephony, when farmers used barbwire fences for transmission and ran networks out of household kitchens. Today's concerns about what happens to local businesses when stores like Walmart show up in small communities replay what the realities were for stores in isolated, small towns when the postal service arrived, carrying catalogs from stores like Montgomery Ward and, literally, delivering the goods. Over and over again, efforts that it was believed would strengthen rural communities actually undermined them; electricity did, as was projected, make farm life easier, but because it also reduced the amount of labor needed, jobs were lost, causing young people to leave for urban opportunities.

These authors provide multiple examples of policy precession, interactions among the effects of various types of policy typically thought of only in siloed terms. Before the post office offered rural free delivery to homes, for example, some had their mail read to them over the telephone by the post office. A mandate for "post roads" is in the US Constitution, but to provide support for the extension of service to rural areas, those in many rural communities pooled their resources to build the roads that the postal service needed for rural free delivery. Policy proposals for government support of professional news organizations to help ensure their survival are in play today; Sawhney and Ekbia note that once rural postal service was in place, local newspapers flourished, new papers being founded where none had existed before and those that had been weeklies or biweeklies becoming dailies.

Sawhney and Ekbia introduce an analytical approach that they call "recentering-on-reversal," a technique that starts by thinking about what is happening at the margins, what they refer to as "the difficult bits," to understand the dynamics in play as they are experienced by those for whom the margins are the center. Because this approach by definition leads to an expansion of the range of perspectives taken into account, it will be useful for any type of policy analysis.

The book necessarily focuses on a single country, the US, for its case studies comparing the treatment of diverse types of systems within one political and legal system, but the approach can be used anywhere. Internationally comparative work would be valuable, for there are both historical and contemporary examples of quite different national approaches to universal service in particular sectors, and there have been reversals. For a very long time, telephony was managed by governments everywhere but in the US, where it was private from the start except for a brief period of nationalization during World War I. In the last decades of the twentieth century, though, countries around the world moved, one way or another, toward the US position, making analysis of this case more useful now than it would have been a few decades ago. Where the US is, as Sawhney and Ekbia note, increasingly considering, and in some states offering, universal access to higher (tertiary) education by making it free, Germany has long provided that, but the UK gave up on it in the 1990s.

Universal service discussions most often focus on individuals, but system nodes can also include households and communities. This complicates our understanding of universal access as a key information policy concept. It is not binary; it can come in different flavors. For those who study infrastructure, the book is a must-read, with its elucidation of commonalities and differences across sociotechnical systems and its literal doubling of the kinds of questions that should be asked of any such system. For a course on sociotechnical systems, the book could valuably be paired with Carolyn Marvin's *When Old Technologies Were New*.

The question of what difference it would make to a community if it had better communication connectivity, asked in recent decades regarding the telephone and the Internet in developing societies, is evergreen.

In every instance discussed here, the answer has been that the extension of universal access has brought economic benefits to both systems and individuals, though not necessarily to communities, and the conclusions are not simple ones. Importantly, though, Sawhney and Ekbia prefer not to use the terms "costs" and "benefits," for more than economics is involved. What does it mean to create an education system devoted to training labor rather than the flourishing of the child? What happens to the distinct pleasures of rural life when it becomes more and more like living in a city?

Over and over again, this book shows how the recruitment of the particular, the idiosyncratic, the genuine local into national, corporate, and industrial systems that is accomplished through universal access can also be experienced as diminishment. In this sense, it can also be read as a paean to "the difficult bits." Hear, hear. This may or may not be the final book from these two authors, but either way, it is great late style.

ACKNOWLEDGMENTS

The scope and complexity of our book project posed many challenges, some expected and others unexpected. We were fortunate to receive support from key people who helped us further our endeavor.

We have to start with Sandra Braman, the editor of MIT Press's Information Policy Series, who saw potential in an unusual idea, first shared with her on Microsoft PowerPoint slides and then developed into a book proposal. Subsequently, she marshaled the review of the book proposal by four reviewers, and that of the completed manuscript by another four reviewers. We can only imagine the difficulties of recruiting so many reviewers from a range of disciplines for a very particular and complex book project. We have to also thank Gita Manaktala, editorial director of MIT Press, for recognizing the potential of our project and supporting its ambition.

We received over 20,000 words worth of comments from anonymous reviewers. Clearly, they invested a lot of time and effort in their reviews. We benefited from their erudite commentaries and the perspectives they brought into play. Since the scope of our book project took us beyond our own areas of research, we also sought scrutiny and guidance for system-specific chapters from scholars with relevant expertise. We were fortunate to secure such help from Richard John, Fred Lerner, Jennifer Lieberman, Milton Mueller, Bob Pepperman Taylor, and Derek Vaillant. We were touched by their generosity, especially of those with whom we had no prior contact, in person or over email – their warm response to our "cold email" was heartening.

Our book project also benefited from the input of two former students. Krishna Jayakar, once Sawhney's doctoral advisee, now a full professor at Pennsylvania State University, provided steadfast critiques that eventually took critical parts of chapter 1 to the place where they needed to be. In the summer of 2019, Andrew Brown, a former undergraduate student in Sawhney's class, emailed for a discussion over coffee. Interestingly, he wanted to discuss readings from the class he had taken last year! Over the course of the conversation, he happened to mention that he had done graphics work for professors. Subsequently, he made professional-quality versions of our figures for this book.

Finally, going beyond people with whom we had occasion to interact one-on-one, we would also like to thank scholars who did research on the systems we cover in this book. Without their extensive scholarship, this book would have been inconceivable.

1

INTRODUCTION

Today, while we may debate the specifics, we widely agree that everybody should have an internet connection (i.e., universal access), since effective participation in our networked society requires it. Some commentators have even characterized the existing shortfalls in universal internet coverage as "shameful" (Mazzocato 2020), an "unfulfilled moral obligation" (Simama 2020), and "an emergency" (vanden Heuvel 2020). Recently, US president Joseph R. Biden advanced a $100 billion plan to "reach 100 percent high-speed broadband coverage," noting that "it is necessary for Americans to do their jobs, to participate equally in school learning, health care, and to stay connected" (White House 2021).

At the time of Bell System's breakup in 1983, about four decades ago, the telephone network was serving 91.4 percent of US households—eclipsing the world at large (FCC 1998). Yet the court-mandated separation of long-distance (toll) service from local service made people in many quarters anxious about universal telephone service because monies from the former were subsidizing the latter. In the fallout of the breakup, the question of the day was: "For whom the Bell toll?" (Nickolai 1984, 507). Subsequently, the stakeholders worked out new arrangements, telephone coverage expanded to around 95 percent of US households, and the 100 percent coverage ideal lost relevance because with technological advances, access to broadband and smartphones became a bigger issue.

Earlier, in interwar years, universalization of electricity was the big issue. By 1932, 70 percent of Americans had residential service nationally, but only 10 percent in rural areas (Meinig 2004). Troubled by the fact that rural dwellers were still laboring "by the sweat of their brows" in the electric age, proponents of rural electrification sought alternatives to the ways of power companies (*Rural Electrification News* 1936b). Believing that the objective of rural electrification projects should be long-term economic sustainability as opposed to profit maximization, they developed innovative ways to make that possible. President Franklin D. Roosevelt, a supporter of these efforts, established in 1935 the Rural Electrification Administration (REA), which provided soft loans at below-market rates and technical expertise to rural electrification cooperatives, expanding electricity in rural areas. At around the same time, another technology spread rapidly—radio—reaching over 95 percent of households by 1950 (Fischer 1992). Unlike electricity, its universalization did not call for policy intervention because people only needed radio sets, which were not expensive, to access broadcasting. In the case of broadcasting, universal access concerns and policy interventions were related to content production and availability—a diversity of voices, as opposed to connectivity.

Further back in history, by 1863 the postal service, which had been expanded by George Washington and his successors, was delivering mail to addressees' doorsteps in urban areas—but not in rural areas, where people had to pick up their mail from a post office, often in a nearby town. Rural dwellers' cries of unfairness were countered by the voices of opponents, who reckoned such a service in rural areas to be economically foolhardy. After much contestation, in 1893, Congress passed legislation for Rural Free Delivery (RFD), which extended delivery to individual rural addresses.

In the nineteenth century, we also saw two other universalization movements—universal education and public libraries. Certainly, equality and fairness were major factors in the expansion of universal education: "All children are entitled to equal education; all adults to equal privileges," as the banner of newspaper *Working Man's Advocate* affirmed in the 1830s (Binder 1974, 33). However, as we will see a little later, universal education in actuality was a product of a confluence of many agendas, articulated and not. Born of a similar confluence of agendas, the public library movement was tied to that of universal education, with its leadership taking libraries to be complements of schools, extending education to students in their after-school hours and adults past their schooling years.

At this point, let's step back and look at all these universalization efforts together. We see that our universal internet access undertaking is not singular, but a chapter in a much longer story—a recurrence of a dynamic that played out a number of times over the last 200+ years.

Yet, while researchers have written extensively about the universalization of these systems, they studied one system at a time. Further, their system-specific works rarely refer to experiences with another system. When they do make connections, they typically do so in a general manner, such as arguing against policy intervention for universal internet access. Compaine (2001b) says that the so-called digital divides are transitory gaps, which will disappear with technological advances and reduction of prices, as was the case with broadcasting. Taking a diametrically opposite stand, arguing for major policy interventions, the Biden administration says that "broadband internet is the new electricity" (White House 2021). Very few studies on the universalization of one system have systematically analyzed the experience with another system, but there have been some; for instance, Sawhney (1993) and Sawhney and Jayakar (1999) analyze the development of universal education in their studies on universal telephone service. In this book, we embark on a systematic analysis of the universalization of all seven systems discussed here.

We are motivated to do so by deeper questions that go beyond universal access per se. Why did universal anything become an important social issue only in the last 200-odd years? Why was it not an issue 2,000 years ago, when prophets roamed the earth? Why did egalitarianism suddenly become desirable? From this perch, we started seeing universal access as a distinctive social phenomenon of our times, as opposed to a remedial, technocratic exercise. We also gained distance from our own commitments to and investments in the issues of the day. We began looking at things in dispassionate and forensic ways, our questions prompting us to go beyond what is said to what is *not* said, beyond what is done to what has *not* been done, and beyond what is measured to what is *not* measured in the realm of universal access.

Billions of dollars are being spent on universal access programs, and even more will be spent in the future, even if President Biden does not get his way with the Congress. The Universal Service Fund of the Federal Communications Commission (FCC) alone, for example, disbursed $8.5 billion in 2020 (Universal Service Administrative Company 2021). There are many other universal access programs in the communications arena, including some sponsored by states, let alone programs in other domains such as education, libraries, and electricity. With our wider perspective, across systems and across time, we are able to identify blind spots in established policies and practices and also broaden our understanding of the issues at play. By doing so, we hope to lay the groundwork for a fuller and more honest discussion on issues related to universal access.

To execute this project, we have employed a particular approach. Next, we explain that approach—our analytical framework and the technique we use to go beyond what is said or done to what is *not* said or done. Subsequently, we explain our selection of cases and outline the theoretical arc that runs through this book.

OUR ANALYTICAL APPROACH

Our framework unfolds from a simple idea—namely, that the beneficiaries of aid also suffer travails, and, conversely, benefactors of aid also gain. We start by spelling out the aforementioned idea, first with a concrete illustrative case—universal education—and then with a typology.

The ideal of universal education already existed when the US was born. In his first address to Congress, George Washington emphasized the need for education. Further, Thomas Jefferson, Benjamin Rush, Noah Webster, and other proponents of education advanced a number of plans for its expansion. Also, the American Philosophical Society held a competition for plans to improve the nation's education system (Cremin 1980; Madsen 1974). In effect, there was no shortage of ideas, but no systematic effort was made to put them into practice.

What was missing was the will to "translate sentiments into appropriations" (Ditzion 1947, 10). Indiana, the first state to inscribe "gratis and equally open to all" education in its constitution, for instance, allowed for a delayed realization of universal education, adding the phrase "as soon as circumstances will permit" (Constitution of the State of Indiana, 1816, Article IX, Section 2). The ideal alone did not muster the political will, and the qualifying clause provided a convenient excuse for delay (Meyer 1965).

Then, in the mid-nineteenth century, the agendas of major players converged in ways that propelled the development of universal education. The industrialists needed a trained workforce. Labor leaders believed that child labor depressed wages in an era of large-scale immigration and surplus labor. The propertied elites, apprehensive of "mob rule" by recently enfranchised propertyless white males, saw universal education as a means of "reconciling freedom and order" by conditioning them to buy into the established system (Kaestle 1983, 5). Ordinary people, fearful of Catholic immigration, were willing to pay taxes for public schools, seeing them as the "principal digestive organ of the body politic" for Americanizing Catholics (Strong 1963, 89).

Within this context of overlapping interests, we start to see the development of the common school—tuition free, tax funded, nondenominational, and universal (Sawhney and Jayakar 1999). To foster its development, legislators had to pass laws for levying taxes, enforcing compulsory attendance, and outlawing child labor (Alexander and Jordan 1973; Edwards and Richey 1963; Kotin and Aikman 1980). Correspondingly, the state created an educational bureaucracy to implement the laws and oversee schooling (Katz 1976).

In sum, disadvantaged children and their families were not the only beneficiaries of universal education. There were others, including the state and the industry. Furthermore, as we will show in chapter 3, the state and the taxpayers were not the only ones to travail—children and their families did as well.[1]

Figure 1.1a captures the essence of this point—a full understanding of universal access requires an examination of the gains and travails of the individual and of the system.

In actuality, as we will show in the chapters of this book about the different systems, our focus largely tends to be on the gains of the individual and the travails of the system (figure 1.1b).

Our argument is that we need to counteract bias and also look at the system's gains and the individual's travails (figure 1.1c).

The idea is to go beyond the perspective of particular stakeholders and lay out the totality of gains and travails in order to further an honest

1.1a Gains and travails of the individual and the system.

1.1b A partial understanding of gains and travails.

1.1c A fuller understanding of gains and travails.

conversation on universal access. For instance, in chapter 3, on education, the main points from our in-depth analysis are summarized in table 1.1. It lays out an outline of an honest conversation on the question: Should children be molded to enhance the well-being of society, or should society should be changed to enhance the flourishing of the children?

To conduct such analysis for each system, our challenge is to go beyond what is said or done to what is *not* said or done. We employ a particular technique to do so—recentering-on-reversal. Here, we build on Agre's (1997) work on performing a reversal on the prevailing intellectual system or mentalité in lay terms. According to Agre (1997, 45), "in the center are those phenomena that are readily assimilated to the system's generative metaphor; in the margin are those phenomenon that can be assimilated to the generative metaphor only by means of unreasonable convolutions." These "difficult bits of intellectual systems" (43) on the margin raise questions about the certitude of the established order.[2] The "systematizers" (43), therefore, work to brush such "difficult bits" under the rug by

Table 1.1 Universal education: Gains and travails

	Gains	Travails
Individual	Increased earning power Skills to function in the world beyond one's immediate environment Social benefits such as improved sanitary practices, family planning, and women's empowerment	State's encroachment on parental control of children Secularized education that does not offend anybody Centralization and bureaucratization of education Prioritization of the needs of state and economy over the idiosyncratic flourishing of children
System	Educated workforce Increased capacity to create and staff complex systems in industry, military, medicine, etc. Savings such as in health care with improved personal hygiene Assimilation of immigrants	Unavoidable trade-offs of the sort that give rise to unhappy constituencies Expansionary tendency—there are always constituencies pushing for upping the level of universal education, to higher grades in school, to community college, to four-year college

explaining them away, making fun of them, and obscuring them. From an analytical standpoint, these "difficult bits" are a window into a different way of looking at things, which the prevailing intellectual system obscures. We should use these as a handle for a reversal—dethroning the center and center staging the margin.

To maintain analytical clarity, note that Agre is not talking about flipping the figure and the ground, as in gestalt psychology. He is also not talking about something akin to Bertolt Brecht's alienation effect—making the familiar strange to denaturalize established narratives and configurations. With regard to specific theorizations on infrastructure networks, he is not even talking about Edwards's notion of infrastructure inversion: "To understand an infrastructure, you have to invert it. You turn it upside down and look at the 'bottom'–the parts you don't normally think about precisely because they have become standard, routine, transparent, invisible" (2010, 20). More broadly, criticism directed at the center from a position on the margin does not constitute reversal by itself. Reversal is a very particular analytical action.

To avoid confusion with other more familiar and often similar-sounding concepts, we use the term "recentering-on-reversal." Furthermore, we delineate the three steps in the performance of a reversal (figure 1.2): One, identify and focus on the difficult bit of an intellectual system, as opposed to papering over or dismissing it (step 1). Two, move the vantage point to it, which is to recenter (step 2). Three, see the world from this new vantage point (step 3). As a result, you can benefit from an expanded understanding—the earlier view plus the new view (see "Expanded Understanding," in figure 1.2).

To us, after several years of research on metaphors, vocabulary associated with reversal comes naturally. If the reader prefers to go beyond the confines of this vocabulary, here is another way of talking about this technique. Signs of the hidden dot the edges of our field of vision, but we gloss over them because they do not fit into our view of things. However, we should deliberately look for them. Furthermore, upon identifying them, we should switch perspectives and understand the alternative view, wherein the hidden is fully visible and normal—such a switching of perspective is key.

It is important to note that the objective here is not to demolish the entrenched metaphor. It is instead to decenter it so as to expand our thinking beyond its boundaries. When we center-stage "difficult bits of intellectual

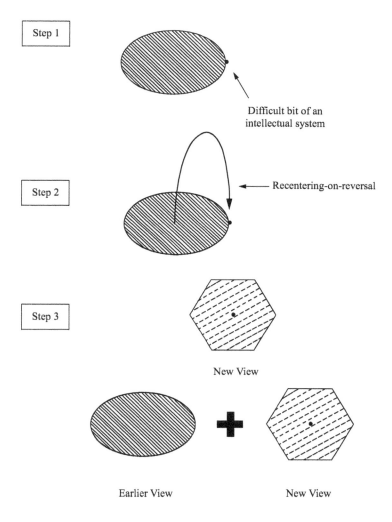

EXPANDED UNDERSTANDING

1.2 Recentering-on-reversal.

systems" at the margins, they open up new vistas, releasing us from what G. K. Chesterton called "the clean, well-lit prison of one idea" (Rosch 2001, 245). This new field of vision is also bound by its core metaphor, but it is different from the old one and complements it, thereby expanding our understanding. In this manner, we apply the analytical strategy of recentering on reversal to foster a broader understanding of universal access.

SELECTION OF CASES

While scholars are in general agreement that case selection in qualitative research should not try to mimic statistical sampling, they are not of one mind on what the selection criteria should be. Broadly, they gravitate toward either theoretical sampling or purposeful sampling (Curtis et al. 2000; Ragin 1992a; Ritchie and Lewis 2003).

In accordance with grounded theory approach, which starts with initial data collection and subsequently engages in continuous interaction between emergent theory and data, Eisenhardt notes that "the goal of theoretical sampling is to choose cases which are likely to replicate or extend the emergent theory" (1989, 537). Here, cases are selected and added over the course of the research in an iterative process guided by emerging theoretical insights. On the other hand, with purposeful sampling, the cases are selected with a clear analytical intent—"strategically selecting information-rich cases to study" (Patton 2015, 265). We employ purposeful sampling, given our specific analytical intent—checking for bias, identifying blind spots, and exploring the utility of recentering-on-reversal.

For purposeful sampling, researchers use a number of strategies, such as homogeneous sampling, typical case sampling, maximum variation sampling, critical case sampling, and confirming and disconfirming case sampling (see Exhibit 5.8 in Patton 2015 for a comprehensive list of forty sampling strategies). Among these options, maximum variation sampling is most appropriate for our analysis because it employs a wide range of cases on dimensions of interest in order to identify shared patterns. According to Patton (2015), "for small samples, a great deal of heterogeneity can be a problem because individual cases are so different from each other. The maximum variation sampling strategy turns that apparent weakness into a strength by applying the following logic: Any common patterns that emerge from great variation are of particular interest and value in capturing the core experiences and central, shared dimensions of a setting or phenomenon" (283). For its implementation, the researcher needs to identify the critical dimensions for the subject in question. For instance, in a study on schools in a state, geographic variation should be represented in the selected cases. Research based on such sampling yields "two kinds of findings: (1) high-quality, detailed descriptions of each case, which are useful for documenting uniqueness, and (2) important shared patterns that cut across cases and derive their significance from having emerged out of heterogeneity" (Patton 2015, 283).

MAXIMUM VARIATION SAMPLING

To implement maximum variation sampling, we:

1. Specify our cases.
2. Construct a sampling frame.
3. Develop a maximum variation typology and select cases.
4. Discuss the number of cases selected.

(1) Specification of cases In their edited volume, Charles Ragin and Howard Becker (1992) set out to answer a question that is also the book's title: *What is a case?* They find that researchers greatly vary in their understanding of what constitutes a case. Ragin (1992b, 9) broadly groups their approaches into four categories. In the first, cases are objects—cases are "empirically real and bounded" (e.g., nation-states). Here, researchers do not have to explain their existence or justify their boundaries. Second, cases are conventions—they are generally agreed upon and "external to any particular research effort" (e.g., industrial societies). Here, researchers do not have to explain their existence or justify their boundaries either. Third, cases are found—they "must be identified and established in the course of the research process." For instance, Douglas Harper (1992), one of the contributors, explains why he delineated "community" differently in two studies: social network of a rural mechanic in one and an area with fifty dairy farms in another. Fourth, cases are made—they are "specific theoretical constructs which coalesce in the course of research" (Ragin 1992b, 10). Michel Wieviorka (1992), another contributor, describes how his understanding of cases changed over the course of his research on terrorists. At first, he examined them from the perspective of established theories, such as functionalist theories of political violence and resource mobilization theory. In the course of his research, he developed the notion of "inversion"—claiming to speak on behalf of a group and, at the same time, distorting its tenets—and thereupon, what constituted a case for his study changed. To add to the fluidity, these categories are not absolute—a researcher could start with a case conceived in one way and move toward another conceptualization during the course of the research; for instance, a researcher could start with a conventionalized case and later develop a new case construct (Ragin 1992b). Acknowledging this fluidity, Ragin concludes by stating two points: One, "the biggest obstacle to clear thinking about 'What is a case?' is the simple fact that the term 'case' is used in so many different ways" (1992a, 217). Two, it is more productive to think in terms of "casing"—"making something into a case" (1992a, 218).

Of all the various ways of casing, we follow Thomas's (2011) approach. He characterizes the delineation of cases on the bases of places, events, and time periods (e.g., Vichy France) alone as inadequate. According to Thomas (2011), while "a subject of study might satisfy conditions of singularity, boundedness, and complexity it would not be a case study . . . unless it could be said to be a case *of* something, and that *"of"* would constitute the study's *analytical framework*" (512, italics in original). In other words, a case is a product of the analytical framework. It should illuminate the strengths and weaknesses of the analytical framework to further its development. He elaborates that "the *subject* [subject of the case study] will be selected because it is an interesting or unusual or revealing example through which the lineaments of the *object* [analytical framework] can be refracted" (Thomas 2011, 514, italics in original).[3]

To illuminate our analytical framework, the subject of a case study needs to satisfy all three of the following criteria:

1. It is a system with sustained internodal connectivity,[4]
2. It is a system that has *already* undergone expansion because of universal access initiatives.
3. It is a system for which universal access initiatives have been undertaken across the US.

It is important to note that we do not engage with the question of whether access to any particular system should be universalized, or, more broadly, questions related to the expansion of citizen and human rights. We only analyze retrospectively what happened when the universalization of a system was undertaken in the US.

(2) Construction of a sample frame   According to Ritchie, Lewis, and Elam (2003), there are two types of sample frames: established frames (existing lists, official statistics, etc.), and specifically generated frames for a study.

Scholarship on universal access tends to focus on a single system (e.g., Brown 1980; Fuller 1964; Kaestle 1983; Mueller 1997). In such works, references to experience with other systems tend to be rare—studies on telecommunications networks at times draw on experiences with the postal system (e.g., John 2010), rural electrification (e.g., Sawhney 1993), and education (e.g., Sawhney 1994; Sawhney and Jayakar 1999), studies on public libraries often discuss universal education (e.g., Garceau 1949; Garrison 1979; Lerner 2009; Nardini 2001). In effect, no master list of systems that have undergone universal access expansion exists.

On the other hand, comprehensive studies on particular locales where many systems coincide exist: "hard" and "soft" infrastructures of cities (e.g., Coutard, Hanley, and Zimmerman 2004; Graham and Marvin 2003; Klinenberg 2018; Tarr 1984), and also specific nodes on this web of infrastructures—the office (e.g., Chilton 2012) and home (e.g., Handlin 1979). Also, comprehensive studies that situate systems in larger social processes such as industrialization (e.g., Beniger 1986) and globalization (e.g., Castells 2011) exist. Drawing on such sources, we compiled our sample frame, shown in table 1.2.

With regard to the adequacy of a researcher-generated sample frame, Ritchie et al. (2003, 88) ask two questions: Does the frame provide the details required to inform selection? Does the frame provide a comprehensive and inclusive basis from which the research sample can be selected?[5] Our frame here is small, and we have looked at the literature on all the systems included. We believe that our efforts have yielded a comprehensive, if not exhaustive, frame.

(3) Maximum variation typology and sample selection We identified two critical characters on which the systems in the sampling fame vary:

- *System assemblage.* Some systems are relatively easy to deploy, and conversely, others are relatively difficult to deploy because of the nature of their assemblage—the "mode of ordering heterogeneous entities so that they work together for a certain time" (Müller 2016, 28). In an assemblage, the type of connectivity that holds the heterogeneous entities together is a critical defining feature. Broadly, it could be of two types: high and low coupling. In an assemblage with high internodal coupling, on isolation, a node is more or less incapacitated (e.g., a local telephone switch). In contrast, in an assemblage with low internodal coupling, on isolation, a node can continue to function, albeit at reduced capacity (e.g., a local school).

Table 1.2 Sample frame

	Universal Access Initiatives Undertaken
Automobility[a]	PARTLY Roads: Yes (federal-, state-, and local-level initiatives) Personal vehicles: No Public transportation: Limited local initiatives
Broadcasting	YES
Education	YES
Electrification	YES
Health care	PARTLY Patient Protection and Affordable Care Act ("Obamacare") significantly expanded coverage, but access is still far from universal.
Internet	YES
Postal system	YES
Public libraries	YES
Sewer system	YES
Telephony	YES
Water supply	YES

Note: a. "Automobility" refers to motor vehicles and highways as a system.

- *Universalizing impulse.* Whoever sets the ball rolling colors the development of universal access policies and programs, from inception to maturation. Broadly, the universalizing impulse could be of two types: establishment initiated and grassroots initiated.

We then developed a typology (figure 1.3). Thereafter, we worked through the sample frame and matched potential cases with the four quadrants of the typology. We settled on cases that had the most relevance for information policy, and they were in seven areas—the postal system, education, electrification[6], telephony, public libraries, broadcasting, and the internet. Among the selected cases, broadcasting is something of an outlier. Unlike other systems, in the case of broadcasting, user connectivity to the broadcasting system has not been an issue, as a user only needs to buy an affordable device (a radio or television set). Instead, the issue has been about programming, and

Universalizing Impulse

	Establishment Initiated	Grassroots Initiated

1.3 Maximum variation typology.

in particular whether it afforded access to a diversity of voices. We saw two benefits in including broadcasting. One, an outlier case "may illuminate the object [analytical framework] by virtue of its difference" from the key cases (Thomas 2011, 514). Two, the elements that mark its "difference" (programming content) resonate with the internet, a multifaceted system, in ways that those of the key cases do not (Lund 2014, 226).

(4) Number of cases: Thickening In our case study research, we have single studies, paired comparisons, and what Klotz (2008) calls "more-than-two and not-a-lot." She notes that while the logic for selection of a single case and a paired comparison can be clearly delineated, there is "no single formula for dealing with multiple cases" (55). At the same time, Stake (1995) notes that "balance and variety are important; opportunity to learn is of primary importance" (6). Our choice of seven cases was motivated by "balance and variety" in service of thickening of understanding, which we explain below.

In 1973, the eminent anthropologist Clifford Geertz (1973) introduced Gilbert Ryle's notions of "thin description" and "thick description" to a broad audience. The former phrase refers to simple factual descriptions in the I-am-a-camera mode, and the latter to "a stratified hierarchy of meaningful structures" that contextualize the situation and bring out relationships between actors, providing a sense of intentionality, history, and potential future (Geertz 1973, 7). Norman Denzin elaborates on the idea of "thick description" in this manner:

A thick description . . . does more than record what a person is doing. It goes beyond mere fact and surface appearances. It presents detail, context, emotion, and the webs of social relationships that join persons to one another. It enacts what it describes. Thick description evokes emotionality and self-feelings. It inserts history into experience. It establishes the significance of an experience or sequence of events for the person or persons in question. In thick description the voices, feelings, actions, and meanings of interacting individuals are heard, made visible (2001, 100).

Over time, many disciplines—sociology, psychology, communication, education, and others—adopted the notion of "thick description." Methodologically, to generate a thick description, they went beyond ethnography, Geertz's prime focus, to interviews, focus groups, and other methodologies (see Ponterotto 2006 for an overview of this evolution). So much so that Denizen (2001), even developed a classification scheme, identifying twelve types of thick descriptions:

- micro thick description
- macrohistorical thick description
- biographical thick description
- biographical-situational thick description
- situational thick description
- relational thick description
- interactional thick description
- inclusive thick description
- incomplete thick description
- glossed thick description
- purely descriptive thick description
- descriptive and interpretive thick description.

Furthermore, scholars have also developed extensions of the concept, such as "thickness" and "thickening strategies."

According to Morrow (2005), "the 'thickness' of the descriptions relates to the multiple layers of culture and context in which the experiences are embedded" (252). Latzko-Toth, Bonneau, and Millette (2017) further extend "thickness" to "thickening." In face of rising big data research, they urge qualitative social media researchers to engage in "'thickening' the data" by adding at "different times through the research, multiple layers—of description, historical and social context, cultural meaning" (202). They go on to offer a three-layer model of data thickening: contextual information (first layer), thick descriptions of practices under study (second layer), and meanings that users attribute to them (third layer). In effect, while many disciplines have embraced the sensibility spurred by the notion of "thick description," the way that they have brought it about has varied.

In our particular endeavor, while our primary focus is on the "shared pattern," we also seek to thicken our understanding by examining the variations from it—that is, "divergences" of each case. In other words, we seek to know not only what sorts of blind spots occur, but also which ones occur in what contexts. We do this by producing gains and travails tables, like table 1.1, for each system, and then synthesizing their contents in chapter 9, "Conclusions." We also conduct a similar exercise for thickening our understanding of how recentering-on-reversal can or cannot help in the identification of blind spots in various contexts.

COMPLETING THE PICTURE: FROM INCLUSION TO ALSO BINDING

In the subsequent chapters of this book, we examine the efforts directed at the universalization of the postal system, education, electrification, telephony, public libraries, broadcasting, and internet. There are entire books on each of the systems we cover. Our book is neither a compilation of their histories nor a conventional comparative study. It is a juxtaposition, put together in order to bring forth conceptual development.

In each domain, we first discuss the initial expansion of the system. We then explore the nature of the universal access intervention and what motivated it. We thereon take stock of the gains and travails of the individual and the system, employing recentering-on-reversal strategy to go beyond what is said or done to what is *not* said or done.

We find that in the development of universal access, the logic of inclusion is plain—articulated and openly implemented, focusing our attention on the individual's gains and system's travails. We identify another logic in operation as well—the logic of binding, which is not articulated and quietly implemented. Consequently, policymakers, researchers, and other stakeholders have not been paying due attention to the system's gains and individual's travails.

The logic of inclusion works through the triad of inclusiveness, access, and benefit. It articulates the following: To include all types of people (regardless of their income, race, age, gender, or other characteristics) in processes and forums critical for equitable participation in society (whether political, social, or economic), the enabling systems need to be universalized. Correspondingly, it calls on large amounts of money to be spent for universal access. And these amounts are indeed large, as noted earlier. Herein, the individual's gains and the system's travails are made visible.

But this is not the complete story, as universal access is not *entirely* humanitarian in character and intent.[7] Any endeavor that entails the movement of huge amounts of money brings forth complex agendas of the stakeholders, who are motivated by a mix of humanitarian concerns and self-interests. The center-staging of the humanitarian piece tends to make the self-interest piece imperceptible.

We show that another logic is also at play, which works through the triad of conscription, conversion, and travail. As this logic goes, universalization of systems, while benefiting individuals, binds them to specific ways of life, behavior, and interaction by converting them to particular ideal types, imposing travails on them in the process. To cite a simple example, the postal service will deliver mail to one's house only if one installs a mailbox of approved dimensions at a specified height—one that aligns with the postal van. Moreover, one is obligated to receive junk mail, which is a major source of revenue for the postal service.[8]

Individuals are conscripted into systems because it enhances systemic benefits. E-filing one's federal taxes, for instance, reduces the costs of collecting taxes, electronic transfers cut down costs of disbursing welfare monies, automated bill payment reduces the transaction costs for corporations. The reach of the market is also expanded. One of the major beneficiaries of the expansion of the postal service in rural areas, for example, were mail-order companies, in the same way that online retailers today have benefited from the expansion of the internet. The reach of the state is also increased. For instance, with conscription, the marginalized become more visible, as they have to register to avail themselves of universal access resources. In effect, individuals are also resources for the systems.

Binding is not necessarily bad. At times, individual decisions do not align with what is socially optimal. Systems often subsidize prices and offer other inducements to individuals in order to attain optimality. The question is how far the system should go when using inducements to secure universal access, especially when "private non-monetized costs may be high" for the individual (Jayakar 2017, 16). This moral question has no clear answer. Perhaps some coercion even could be justified—this deeper issue also has no clear answer. Such decisions should be made democratically, and as transparently as possible. We seek to map the gains and travails of both the system and the individual to provide fully informed grounds for what are ultimately moral decisions.

2

POSTAL SYSTEM

In 1899, Alaska's vast interior—"the last frontier"—had no mail service. Eight years later, in 1907, it had mail delivery routes, ranging from 150 to 650 miles long. The frequency of delivery, though, was low—some routes had a reasonable, twice-a-week service, some only had deliveries twice a year, and others fell somewhere in between. The carriage was by dog-drawn sleds, reindeer (occasionally), and "half-breed runners." After Alaska converted its trails to hardened roads, the post office pressed horses into service, enabling more frequent conveyances of heavier mail (*New York Times* 1909a). This mail service was part of Rural Free Delivery (RFD), a universal access project. One *New York Times* article, after describing the arduous delivery routes and the endurance of determined mail carriers and their beasts, concluded: "The Alaskan mail service is one of the government's practical philanthropies. It doesn't pay, of course, in dollars and cents. It pays 32 cents the pound and costs something like $5, but in what it contributes toward making life under particularly desolate circumstances approximately worth living, it pays a hundredfold" (*New York Times* 1909a, 7).

These three sentences express, with arresting visibility, the bias of the entrenched framework—the system's travails ("the Government's practical philanthropies") and the individual's gains ("making life . . . approximately worth living"). With regard to the system's gains, the article briefly talks about publishers protesting the post office's decision, on logistical grounds, not to deliver magazines to northern Alaska; clearly, they saw themselves gaining from the expansion of the mail service. The article also touches on the individual's travail—the price of postage, while also noting that it is low relative to the actual cost of delivery. But it does not consider other possible gains and travails, including nonfinancial ones, to the individual or the system.

We see all the elements of a balanced analytic framework in play—the individual's gains, the individual's travails, the system's gains, and the system's travails. The problem, however, is proportionality, as the account is skewed to the individual's gains and the system's travails. What if the *New York Times* had center-staged the rather curious complaint of magazine publishers? Would the story have been different?

Well, it did not do so, but *we* will—for us, operating on a conceptual level, early 1900s Alaskan mail will be one instance of a bigger phenomenon, if an extreme one. As the *New York Times* (1909a) put it, "The Alaskan mail service is the limit of the rural free delivery idea. It is rural free delivery carried to the logical—or rather to its heroically illogical—conclusion" (7). To see how and why, we need to briefly examine the history of the postal system.

EXPANSION OF THE POSTAL SYSTEM

We can trace the origins of the postal service to the organized systems of document carriage of the ancient empires of Egypt, India, and China—postal routes with stations for resting or switching horses or carriers. What is notable for our present analysis is that the state typically restricted such systems to official messages—that is, citizens could not use them for personal or commercial purposes (John 1995)—a point to keep in mind when thinking about universal access as a modern phenomenon.

The first modern protosystem came into being in 1505, when the Hapsburg emperor Maximilian I granted the Taxis, an Italian family firm with multinational interests, a monopoly for mail carriage. This arrangement proved to be a source of significant revenue for the emperor, as well as profitable for the Taxis, as it also provided service to the public for a fee.[1] In 1614, Prussia took a more direct approach by establishing a state-run postal monopoly, which became the model globally (Noam 1992). Colonial America had a British version of this model, which, after independence, the US continued until Congress passed the Post Office Act of 1792—a radical move, as it emphasized system expansion rather than raising revenue for the treasury.[2] To resource this expansion, it directed monies from profitable routes to unprofitable areas, which "hastened the rapid extension of the mail into the hinterland and the large-scale convey-ance through the mail of newspapers and magazines, then the principal source of time-sensitive information on public affairs" (John 2010, 19).

Among other things, the Post Office Act moved the power to designate postal routes from the executive to the legislative branch—a seemingly minor administrative matter that turned out to be a decisive move toward system expansion. Members of Congress started clamoring for the exten-sion of the postal system to their districts and the subsidies to make that possible.[3] This intense pursuit of self-interest by the members of Congress and their constituents shattered the hold of the earlier logic—namely, that each route should be self-sustaining. The ensuing politics gave rise to a complex system of subsidies from profitable New England and Mid-Atlantic states to the rest of the country (John 1995). In sum, "prior to 1792, the expansion of the postal network remained constrained by the assumption that every new route should be self supporting. After 1792, however, this constraint no longer applied" (John 1995, 49). With this change in direction, the postal service was now in a universalizing mode.

INEQUITIES AND INEQUITIES THAT MATTER

The road to system expansion was not straightforward, however. Prior to 1863, mail was only delivered from post office to post office, with citizens picking up their mail or paying private companies to do so. In some cit-ies, the postal service also delivered mail to the doorstep for an extra fee. Later, in 1863, it started delivering mail to individual addresses in urban areas at no extra charge.[4]

Against this backdrop, rural residents started demanding a similar ser-vice, as they had to travel much farther than their urban counterparts.[5] As a farmer tersely put it: "Why should the cities have fancy mail service and the old colonial system still prevail in the country districts?"[6] They found the disparity galling because they paid the same postage as their urban counterparts.[7]

In this debate, the proponents saw rural free delivery as a matter of fairness, and also as a solution to the problems plaguing rural America. The opponents, on the other hand, considered it foolhardy—huge costs in exchange for only small benefits, in their view.

Amid this debate, release of the 1890 census report, showing a depopulating rural America, brought public attention to several questions that upset the proponents. Why were rural people leaving for the cities? What were rural communities lacking? The discourse centered on the shortfalls of rural life vis-à-vis urban living: "THE POPULAR IMAGE of country life at the turn of the century was a negative one. The causes of rural decline—real and imagined—were typically identified as the social isolation of farm families, urban growth and development, and the general unavailability of modern conveniences in the country" (Atwood 1986, 264, capitalized in original). This understanding of rural problems brought forth a cry for rural-urban equity in all matters, including rural free delivery (Atwood 1986; Fuller 1964; Roper 1917).

To John Wanamaker, postmaster general (1889–1893), it made more sense for one person to deliver mail to fifty people than have fifty people come to town to pick up their mail. Moreover, in his judgment, free delivery to rural homes would increase correspondence; expand markets; boost circulation of newspapers, magazines, and information generally; and reduce isolation, which was critical for retaining youth in rural areas.[8] However, when Wanamaker proposed rural free delivery in 1891, he was laughed at: "The idea of sending a mail carrier trudging over the hill and dale to find some farmer to whom a letter might occasionally be directed seemed ridiculous" (Fuller 1964, 19).[9] While critics could not see how the high costs could be justified for whatever benefits were to be gained from an expanded service, Wanamaker argued that this service would pay for itself (USPS Historian 2005). Further, rural small businesses, which would face increased competition, and intermediaries such as wholesalers, who would be bypassed, were suspicious of Wanamaker's motives because he was a retailing magnate, whose catalog mail-order business stood to gain from rural free delivery (Kielbowicz 1994a; Peck 1972).

Finally, in 1893, Congress passed legislation mandating mail delivery to individual addresses in rural areas and allocated $40,000 to fund its start. However, Wanamaker's successor, Wilson Bissell, postmaster general (1893–1895), refused to implement the new service—RFD—which eventually led to his resignation.[10] His successor, William Wilson, postmaster general (1895–1897), also found RFD "impractical," but he was willing "to make the experiment by the best tests I can devise" (Law 1996, 102). Under his administration, RFD was inaugurated on October 1, 1896, with an experimental service on three routes in West Virginia (Roper 1917; Pope n.d.b). Noting the increase in mail handled, his successor James Gary, postmaster general (1897–1898), observed in his 1897 annual report: "the continuance of rural postal delivery 'will elevate the standard of intelligence and promote the welfare of people'"[11] (Postmaster General 1897, 13).

Still, the question regarding the financial justifiability of RFD persisted. The post office instituted a number of experimental projects, and while they invariably indicated increasing demand, a definitive determination as to whether the demand was sufficient to justify the cost remained elusive. Of the experiments conducted before Congress passed the RFD legislation, Wanamaker noted in his 1892 annual report:

The experiments have mainly related to villages, to be sure, but it has been a daily service and it has cleared a profit. It is easy enough, therefore, to say that the free delivery can be inexpensively extended further and further; and it ought to be done, whether it pays a profit to the Department or not. As the frequency of the deliveries increases, and the service seems to pay its way, if a daily visit is too expensive, let the service begin with a triweekly or a semiweekly service; only let it begin. (Postmaster-General 1892, 12–13)

What is notable here is Wanamaker's argument: rural service should be expanded "whether it pays a profit to the Department or not." Whether a service is deemed to be profitable, however, depends on how one calculates the costs of providing it. For instance, where the post office implemented experimental free delivery, rural residents initially showed considerable enthusiasm, and postal revenue increased. However, after the initial enthusiasm wore off, people started returning to their old pattern of visiting post offices, and revenue growth dropped to the prior years' normal levels. Consequently, First Assistant Postmaster General Frank Jones noted in his 1893 annual report: "In contemplating an innovation of this kind, involving so great an expense, the demands of the business public and the benefits to the people alone should control. It should not be extended merely because the free delivery in cities is maintained at the expense of the general public" (First Assistant Postmaster-General 1893, 55). A few years later, in 1899, however, First Assistant Postmaster General Perry Heath made a "vigorous plea" for RFD, noting that the then-existing 383 RFD service areas had seen increased postal receipts, increased land values, improved roads, better market information and greater profits for farmers, and increased educational possibilities (*Grand Rapids Herald* 1899, 1).

In this muddled way, by the end of 1900, about 2,600 RFD routes were created, serving approximately 1.8 million rural people. While the annual increase in postal revenue in non-RFD areas was only 2.49 percent, in Maryland's Carroll County, full RFD coverage resulted in an increase of 10 percent, leading Postmaster General Charles Smith to estimate a nationwide increase of 7.51 percent if RFD were extended to all rural communities (*New York Times* 1900b).

This back-and-forth of positive and negative assessments notwithstanding, the development of RFD proceeded, albeit unevenly. The underlying reality was that latent demand for RFD existed. For instance, many farmers were paying private carriers to fetch their mail from the post office, and, when telephony became available, many farmers made arrangements with the postmaster to have their mail read to them over the telephone (*New York Times* 1901b; Roper 1917). After the start of RFD, many communities even pooled monies to improve roads to make their communities attractive for the new service[12] (Law 1996).

Two decades after its inauguration as an experimental service, RFD was servicing 42,000 routes, at an annual cost of nearly $52 million (Roper 1917). Whether RFD was worth the cost, which continued to be a point of debate, it was a "new pulsation" in rural America (Postmaster General Charles Smith, quoted in *Grand Forks Daily Herald* 1900, 4).

RFD also served as a springboard for the development of nationwide parcel post, which was proving to be a difficult endeavor in the US. Germany established parcel post as early as 1874, and many European countries followed suit, including Britain in 1883 (Jones 1914). In the US, fierce opposition from powerful interest groups stalled the introduction of parcel post for over two decades; particularly, express companies, wholesalers and other intermediaries, and small-town businesses. Express companies used railroads and other transportation to move customers' packages at about twice the speed of regular freight, and they stood to lose business to parcel post. American Express, Adams Express, and Wells, Fargo & Company, which dominated the business, were very influential politically. Intermediaries such as wholesalers also stood to lose with the introduction of parcel post. For instance, according to an estimate, in 1911, farmers pocketed only $6 billion of the $13 billion that consumers paid for their crops (i.e., $7 billion went into the distribution channel). Conversely, the profits of intermediaries in the movement of products from urban businesses to rural stores increased purchase prices for the farmers. Finally, small-town businesses feared competition from urban mail-order companies. Their associations (e.g., National Association of Retail Grocers, Illinois Retail Merchants' Association) lobbied hard against the introduction of parcel post (Fuller 1964). Leveraging their political influence, these interest groups blocked legislative efforts to introduce parcel post for over two decades (*Cincinnati Enquirer* 1907; Fuller 1964).

Within this larger context, rural areas afforded conducive conditions for the advancement of parcel post. When rural residents purchased something weighing more than four pounds, the maximum allowed by RFD, from a mail-order company or other business, they had little choice but to use express companies. However, the express companies only delivered the package to the nearest town. For the conveyance of the package "the last mile" to their homes, they asked their RFD carrier to bring it from the town (*Chicago Daily Tribune* 1900b, 1904). The post office permitted such arrangements, as RFD wagons had extra space and carriers were afforded opportunities to earn extra money. In effect, rural areas fortuitously developed a parcel post of sorts.

Later, in 1907, implementing a new law that Congress enacted under lobbying pressure, the post office greatly constrained when carriers could enter into package delivery arrangements with rural residents, which effectively killed off the protoparcel post that had developed in rural areas. This turned out to be a boon for the development of nationwide parcel post. On the one hand, it energized the movement for parcel post, and, on the other, the rural experience provided a concrete example of the viability and benefits of the service (Kielbowicz 1994a). As Fuller (1964, 200) notes, "Had it not been rural free delivery, it is likely there would have been no parcel post."

In 1907, Postmaster General George von Lengerke Meyer recommended a rural parcel post, limited to rural districts (John 2012). Critics cried foul, characterizing Meyer's plan as a "entering wedge" for a nationwide parcel post (Miles 2012). Proponents pointed out that the US was already way behind European countries (Cowles 2012). Eventually, in 1911, the proponents of parcel post—big-city merchants, National Grange, Farmer's Union, Farmer's National Congress, and also some local retail and trade associations such as the American Florists Association—prevailed, and the legislation was finally introduced on the floor and passed the next year (Pope n.d.a). Thereby, in 1912, the US became the last industrialized country to have parcel post, indicative of the level of opposition it faced, as it was "arguably the deepest federal thrust into public enterprise in the early 1900s" (Kielbowicz 1994a, 150).

People's response to parcel post was enthusiastic. On January 1, 1913, nationwide parcel post started, and by June 30, the service had sent out over 300 million parcels—a number that the post office characterized as "phenomenal" in its 1912–1913 annual report. Specifically, with regard to universal access, parcel post reached twenty million people who were not within express companies' service areas (Postmaster General 1914, 10).

GAINS AND TRAVAILS

In 1905, a two-paragraph article in the *New York Times* with a notable title—"Fifty New Daily Papers. Rural Free Delivery Responsible for the Demand in Iowa"—reported:

Fourth Assistant Postmaster General De Graw today made public a communication to his department, which states that fifty weekly and semi-weekly papers in towns and cities of Iowa are to be made dailies, backed by an organization of capitalists.

Two pages of the papers are to be printed at one central place and distributed to the towns, where the remaining two pages will be filled with the local news. The plan, it is declared, has been made feasible through the greater demand for dailies by rural residents whose mail is now delivered daily through the rural free delivery service. (*New York Times* 1905, 1)

We see here gains (increased demand for newspapers and new availability of dailies) and also travails (centralization, loss of some local autonomy, and greater sway of capital in community life). Furthermore, we see that RFD did not simply extend mail service—it also enabled socioeconomic realignments in complex and often unanticipated ways, which we will examine systematically in the following sections (also see table 2.1).

INDIVIDUAL GAINS

Beyond the convenience of home delivery, RFD brought many other benefits for rural people. It increased farmers' access to crop prices (*Colman's Rural World* 1900; Morrow 1909; *New York Times* 1900c). It also motivated local governments to improve roads, as the post office would extend service to an area only when the road conditions were adequate (Fuller 1972; *Grand Rapids Press* 1916; *Indianapolis Star* 1909). The postal service's use of automobiles for mail delivery, starting experimentally in 1902, prompted local governments to improve roads further (*Chicago Daily Tribune* 1900a; Cushing 1893; *Kalamazoo Gazette* 1908; Law 1996; *Michigan Farmer* 1900; Roper 1917; Seely 1986). In conjunction with better connectivity, the

Table 2.1 RFD: Gains and travails

	Gains	Travails
Individual	Convenience of home delivery Increased access to crop prices, weather reports, educational, and other information resources Improved road connectivity Increased value of land	Exposure of local small businesses to competition from urban-based corporations System's imperative to install standardized mailboxes, assume snow removal responsibilities, etc. Refashioned local environment with the imposition of a grid-based addressing system
System	Increased sale of newspapers and periodicals Increased reach of advertisers Increased business for mail-order companies Enhanced administrative and security capabilities of the state	Conflict between postal service and publishers

value of rural land increased (*Colman's Rural World* 1900; *Grand Rapids Herald* 1899; Law 1996; Morrow 1909).[13] Further, RFD benefited education (*Colman's Rural World* 1900; *Grand Rapids Herald* 1899) and extended the reach of public libraries (*New York Times* 1902f). Interestingly, RFD carriers also became "peripatetic Weather Bureaus," carrying color signals attached to their wagons[14,15] (*New York Times* 1901d).

Travails   Amid the celebration of farmers' gains from RFD, rural people's travails were not registered. In actuality, however, they were conscripted and made part of the system. To understand this, we first look at how rural people were aligned with the system, and then how their environment was refashioned to facilitate the functioning of the system.

Let us start with the mailboxes. The post office asked rural people to install mail receptacles at the end of their driveways, seeking to eliminate the mail carrier's trek to the front door. But receipt of mail that would have been delivered anyway was not an adequate motivator for them. So, to make it worthwhile for them to invest in a mail receptacle, the postal office also included mail pickup in the service, after a study determined it would not take too much of the carrier's time (Cushing 1893).

People started using all kinds of containers as mail receptacles—milk cans, coal cans, lard pails, soap boxes, apple boxes, and others. The post office, however, needed standardization to promote efficient system functioning. It granted concessions to fourteen companies for the sale of mailboxes of approved designs. Rural people now had to purchase a post office–approved box and install it "buggy high" to receive mail (*Cincinnati Enquirer* 1905; *Morning Herald* 1902; Smithsonian National Postal Museum n.d., n.p.). Some members of Congress expressed deep concern about this arrangement and called for a congressional investigation into the "fourteen boxmakers' trust" (*New York Times* 1901c). As a result of this outcry, the post office opened the mailbox market to any manufacturer, so long as the mailbox was made to its specifications (*Morning Herald* 1902; *Morning Oregonian* 1903; Smithsonian National Postal Museum n.d.). Further, it repealed the requirement that all the patrons along a route have the same type of mailbox. They could now choose from any of the fourteen approved styles. But the carrier had to carry fourteen keys instead of one (*New York Times* 1901a). Here, we see an interesting instance of individual-system negotiation, after the conscription of rural people into the system had been accomplished.

Furthermore, rural people had to "meet the department half way" and remove snow from the road (*New York Times* 1902d). Some would do so promptly, while others would wait until the mail arrived, indicating that the road was usable now. While this strategy conserved effort at the individual level, it created problems at the system level—the carriers' movement was hampered and route completion time increased (*New York Times* 1955).

On another front, when the postal service started parcel post, it needed greater cooperation from rural people in the preparation of packages and labeling (Roper 1917). It started campaigns to enlist the help of newspapers and civic organizations, to teach them packaging—an exercise in instilling new discipline to aid system functioning (Pope n.d.c). In a similar vein, they were coached to mail Christmas gifts in advance and mark on the packages "Not to be opened until Christmas." Furthermore, they were tutored to consult train schedules before mailing perishable materials to forestall their getting held up in the post office (Roper 1917).

Not only were rural people conscripted into the system, but their environment was also heavily reshaped to streamline the functioning of the system. Carey (1989, 225) provides a profound insight into the systemization impulse in modernity—"grid is the geometry of empire."[16] Writing about the creation of time zones, he used the term "empire" to refer to systemization across space. Interestingly, this also held true for modern empires (British, French, and others)—for instance, their practice of sending survey parties before their expansion into new areas. In our case, too, we see it play out in the development of RFD.

Roper (1917) articulates the systemization imperative as follows: "Uniformity is the first principle of postal efficiency" (192). Accordingly, with the introduction of RFD, the post office overlaid, if not erased, local idiosyncrasies in addressing, with roads "named, measured, and blocked off" (Cushing 1893, 1011), and buildings numbered in a systematic fashion (Columbus Ledger 1905). As Cushing (1893) describes it:

Each mile is divided into ten imaginary blocks having a frontage on each side of the road of exactly one tenth of a mile each. Two numbers are assigned to each block, the odd ones on the left and the even ones on the right. Wherever country houses are near enough to be situated within the same block they all have the same block number, but are distinguished by small letters, thus: 246, 246a, 246b, etc. Not only the exact location but the correct distance of every house entrance from some convenient point of departure, say the county seat, can be quickly estimated by dividing the block number by two (there being two numbers to each block for each side of the road) and pointing off one decimal place. For instance, No. 246 is 123 tenths miles, or 12.3 miles from the point of reckoning; or if 246 represents the difference between two block numbers, then 12.3 stands for the distance between them expressed in miles. (1010)

This grid facilitated the movement of nonlocal post office employees through the area to perform system functions—delivery and pickup. Conversely, it lessened the importance of local knowledge, which earlier had been necessary for navigating local idiosyncrasies, such as nonsequential numbering of houses 1 (grandparents), 3 (grandchild), and 2 (parents), where the organizing logic was temporal and not spatial.

Concomitant to the grid, the post office introduced uniforms for the carriers (*Aberdeen Daily News* 1915; *Daily Picayune* 1908). Unlike earlier carriers, who typically were known to the community for a long time, the new ones could be anyone from the system—uniforms signaled and legitimized these outsiders.

This integration occurred at multiple levels, from the postal system to the overall economy. While the gains were material, the travails were profound: rural America became an extension of urban America, as opposed to a special place that the proponents sought to preserve. This was compellingly captured in a *New York Times* piece reflecting on a retreat of the unnamed author to a farmhouse to escape from city life (1907). For this author, the reality turned out to be quite different, with the host wearing ready-made clothes, cooking from magazine recipes, and listening to phonograph records. Likewise, the farmhouse had machine-made furniture, mass-produced bedcovers, and installment-purchased encyclopedias, and everything was "up to date." In sum, rural life had become "a hideous imitation of life in a Brooklyn flat, plus a wider landscape and minus a few conveniences" (*New York Times* 1907, 41).

Furthermore, RFD led to the closure of many small rural post offices—26,000 by 1915 (Atwood 1986; Fuller 1972; *New York Times* 1902d; Peck 1972). Beyond the pain suffered by the affected postal employees, these closures had a major impact on local businesses. When rural residents picked up mail at a rural post office, they also usually made purchases at nearby businesses.[17] With the advent of RFD, though, farmers had fewer reasons to make regular trips to the business area of a town, reducing custom for local businesses. Suffering this squeeze, merchants of Newton Falls, Ohio, sent a petition to Washington, DC, saying that the newly started RFD be discontinued:

This is one feature of the new system which apparently has not been considered. Formerly scores of farmers would drive to the town each day from all localities to get their mail and of course while there would do more or less trading with the merchants. Now they have no occasion to make daily trips to the postoffice [sic] but instead the farmers do their trading perhaps on but one day in the week, and then take a longer drive to the largest nearby city. (*New York Times* 1901e, 3)

In 1902, one rural store owner, a former local postmaster, reported reduction in business by half, as his customers were ordering from department stores via postcards or on the telephone.[18] Characterizing RFD as a "sort of distributing agency of your big department stores," he went on to ask: "How can I compete with the huge operations of such stores?" (*New York Times* 1902a, 5). To add insult to injury, to support his family, he was now serving as the local deliveryman for urban stores.

Such negative impacts on small businesses were the "effects not foreseen" (*New York Times* 1902a). In response, rural proponents launched campaigns to urge rural residents to "trade at home" and sustain local businesses. The editor of *Wellman Advance* asked: "Who sympathized with you when you were sick? Was it your home merchant or was it Sears, Roebuck and Co.? . . . When you want to raise money for some needy person in town do you write to the 'Fair' store in Chicago or do you go to your home merchant?" (Atwood 1986, 270). Many country stores also started to match the catalog prices of mail-order companies, touting at the same time that, unlike with mail-order purchases, the customer can see the product in person (Atwood 1986). Rural people often viewed the sudden drop in prices at their country store with suspicion, as it confirmed their view that local merchants had been overcharging them for years. While local businesses made appeals to them, saying that their closure would kill the local community, such efforts could not stop their decline (Atwood 1986; Kielbowicz 1994b).

SYSTEM

Gains The sales of newspapers and periodicals, and correspondingly the reach of advertising by merchants, increased thanks to the post office's support. The post office allowed its carriers to serve as independent newsagents and sell newspapers and periodicals or accept and collect subscriptions on behalf of the publishers (*Plain Dealer* 1904). On the advertising side, earlier small-town merchants made informal arrangements with postmasters for mailing lists of rural patrons. After the advent of RFD, however, the post office started supplying mailing lists in a systematic way (*Dallas Morning News* 1902; *New York Times* 1902d). Post offices also carried unsold back copies of periodicals returned by local sellers at second-class rates (*New York Times* 1900a).

RFD also benefited commerce, especially after the introduction of parcel post. People enthusiastically took to parcel post, as discussed earlier. Equally important, if not more so, for our analytical purposes: "As some had predicted (or warned, depending on viewpoint), large mailing houses did profit from this increase in reach, especially into the households of rural Americans. Sears, Roebuck and Company handled five times as many orders in 1913 as they had the year before. By 1918 they had doubled their revenues" (Pope n.d.c). Essentially, Sears was already benefiting from RFD;[19] the introduction of parcel post, which carried heavier packages, amplified these benefits.[20] Overall, in Roper's (1917) assessment, the mail-order industry was the biggest beneficiary of parcel post. In effect, RFD was not just a welfare program of sorts, with urban areas subsidizing the rural areas—the former were also major beneficiaries.

Moreover, the state also benefited (*Colman's Rural World* 1900). John (1995, 7) pithily captures the administrative utility of a universalized postal system: "Just as the tolling of the village bell introduced the peasantry of Europe to the linear logic of the mechanical clock, so the regular arrival of the mail extended this linear logic to the farthest reaches of the trans-Appalachian West." John is not just being metaphoric here; in fact, late medieval European governments actively sought to spread public clocks (aka universalization). For instance, around 1473, the Duke of Milan pushed to have the distance between public clocks in his jurisdiction to be 5 kilometers or less (Dohrn-van Rossum 1996).

RFD also had a national security dimension. For instance, in the case of Alaskan mail, when studying a potential 450-mile Valdez-to-Eagle military trail, the War Department included an agent of the post office in its expeditionary party. Consequently, soon after the trail's development, the post office was able to start mail service (*New York Times* 1909a). More generally, American governments have seen the postal service as a critical means of holding the nation together (John 1995; Roper 1917). George Washington himself pressed for improved connectivity with the western lands to forestall their slippage into the British or Spanish folds (Harlow 1926).

Travails Both newspapers and periodicals benefited from the privilege of using second-class mail, which was highly subsidized to facilitate the circulation of information and promote the "educational and moral advancement of the people" (Postmaster General Frank Hitchcock, quoted in *Idaho Daily Statesman* 1910, 1).[21] The subsidies were hefty and consequential. For instance, in 1907, the post office charged about 1 cent per pound to deliver second-class mail when its transportation and handling costs averaged 9.23 cents. Consequentially, given this low rate, in 1907, second-class mail constituted 63 percent of all domestic mail by weight, but it provided only 5 percent of postal revenue (*New York Times* 1909b).

Facing huge deficits, the post office sought to increase the rates for magazines, which constituted 20 percent of second-class mail—costing 5 cents a pound (*New York Times* 1909b). The post office was okay with delivering newspapers (mostly local or regional), which cost 2 cents a pound and were carried for shorter distances than the magazines, especially the national ones[22] (*New York Times* 1909b, 1910b; Post Office Department 1910). Moreover, the magazines carried a high ratio of advertisements to news and educational materials. As James Britt, a senior postal service officer, pointed out: "The magazines are making money not through their educational work but through their advertising. They are making much more money, and it is costing the Government much more to carry them" (*New York Times* 1911a, 3). While magazines were a problem for the postal service as a whole, they were especially so for RFD with its higher transportation costs[23] (*New York Times* 1910c). Postmaster General Frank Hitchcock underscored RFD's contribution to increased circulation of magazines, noting, "Were it not for the rural free delivery service, the circulation of the magazines would be materially reduced" (*Washington Post* 1910, 1).

Hitchcock provided statistics such as the following to the congressional commission on second-class mail rates: *Everybody's Magazine*—the advertising rate increased from $1 per line and $150 per page in 1899 to $2.23 (black-and-white) and $2.50 (colored) per line and $500 per page in 1910; the *Saturday Evening Post*—the advertising rate increased from 25 cents per line and $800 for a page insertion in 1898 to $6 per line and $3,000 for a page insertion in 1910. On the other hand, William Rosenbaum, the president of National Cloak and Suit Company, a major mail-order business, argued that advertisements in high-end magazines generated so many letter and catalog mailings that the post office could afford to distribute them for free (*New York Times* 1911b). Furthermore, the publishers disputed the post office's calculations and put forward their own (*New York Times* 1911a, 1924a).

The peculiar economics of communications systems, with their high fixed costs and low marginal costs, gave latitude for raising these claims and counterclaims (see, for example, *Chicago Daily Tribune* 1905).[24] This led to a protracted conflict.[25]

WHAT WAS MISSED

In the public discourse, RFD was widely characterized as a service to benefit farmers.[26] While the farmers indeed secured many gains, they also weathered travails, which are overlooked. They included conscription into the system (e.g., the purchase and installation of specified mailboxes) and the rationalization of their lived environments (e.g., the systemization of their addresses), all of which were costs of tighter integration into the metropolitan economy.

The integration was system-centric, with a decidedly metropolitan bias.[27] To the credit of rural small businesses, they were alert from the very outset to the dangers of better connectivity with the metropolitan economy. But a coalition of large department stores, newspapers, magazines, railroad companies, and farmers worked against them. The small businesses managed to delay parcel post for over two decades, but they eventually lost out.

The opponents of RFD underscored the costs to the system, often citing dollar amounts. But their figures were invariably disputed, as high fixed costs and low marginal costs of communications systems make any allocation of costs only as good as the assumptions on which it is based. And there is rarely ever a consensus on assumptions.

When all was said and done, the proponents lost out on a fundamental level—the preservation of rural communities. As Atwood notes (1986, 272), "ironically, the problems the mail-order businesses brought to the country—depopulation and rural decline—were the very problems rural free delivery was supposed to solve." RFD intensified rural exposure to consumer culture and also opened new channels for farmers to procure goods, decimating rural small businesses in the process. The former profoundly changed farmers' attitudes and way of life, and the latter undermined the economic foundations of rural communities. Perhaps the problem was the unit of analysis—the farmer. Perhaps the proponents should have focused on the rural community. In other words, they should have sought to preserve the rural community, not farmers per se.

Small business associations and local governments did point out that small businesses were critical for the vitality of rural communities. For instance, arguing against the Parcels Post Bill, supported by the Postal Progress League, Charles Hutsinpillar, a small-town Ohio merchant, told the annual convention of the Ohio Hardware Association: "The effect of such a law is beyond the calculation of man, and almost beyond his contemplation. It certainly would revolutionize the whole system of business, and would be the death knell of the retail houses generally, and in a measure the depopulating of towns and villages, the natural effect of destroying the retail business" (Hutsinpillar 2012, 262). But the rural residents saw such arguments as self-serving. While they appreciated the personal touch of local businesses, their financial contribution to community events and associations, their willingness to extend credit, and so on, they felt that these businesses had long abused their local monopolies and overcharged their customers. The small businesses, for their part, felt that the low rate of turnover in rural communities compelled them to maintain high margins. Furthermore, they served as the terminal point in a complex distribution system, wherein each intermediary had a markup that added up to higher prices for rural customers (Atwood 1986; Kielbowicz 1994b). Here, interdependency bred resentment, as is often the case in human affairs. In effect, the rural community could not cohere as a unit.

Together, in a fragmented way, rural stakeholders also had a nuanced view of connectivity. On the one hand, proponents celebrated the removal of farmers' isolation, and on the other, opponents raised alarm about the negative consequences of connectivity, especially for rural small businesses. Consequently, instead of taking connectivity to be necessarily good, a commonplace presumption in modern times, and a binary condition—connected or disconnected—rural stakeholders discussed various types of connectivity. Initially, the post office intentionally limited RFD connectivity from the local post office to rural houses it served, so that only local businesses could benefit from the new service. To wit, this limited connectivity was meant to strengthen the relationship between local businesses and rural customers, bringing efficiency and convenience without disrupting the local social fabric (*New York Times* 1904, 1908b). The metropolitan bias, on the other hand, came into play when the system was scaled up—made translocal, with RFD connectivity extending beyond the local post office to the rest of the nation. Here, connectivity to the railroad system was privileged, as seen in the following news report: "The experience of the department shows that rural free delivery service from a fourth-class office supplied by a star route is rarely successful, and then only under most favorable circumstances. It has been suggested, therefore, that petitions for service be presented only from Post Offices located directly on or very near a railroad, and having ample railway mail service" (*New York Times* 1902d, 27). There was, in fact, no debate on orienting the RFD system toward the railroads, which materially strengthened the metropolitan bias.

When it came to parcel post, the debate on connectivity was much more vigorous and had many more participants, as prospects for service in rural areas were intertwined with prospects for a nationwide service. For instance, at first, the post office thought of limiting parcel post to RFD routes (i.e., from the local post office to addresses on its RFD routes). Advocates of this arrangement, including President Theodore Roosevelt, believed that it would strengthen the relationship between local businesses and rural consumers, on the one hand, and keep distant mail-order companies out, on the other—keeping the rural social fabric intact. Ironically, both small businesses and rural consumers were opposed to it. Small businesses saw it as an "entering wedge" for opening the way for a nation-wide parcel post. Also, in the short run, they felt that city-based businesses would get around "limited parcel post" by setting up local agents, who could avail of RFD service. On the other hand, rural consumers did not want their choices restricted by a limited parcel service (Kielbowicz 1994b, 87). In this fashion, parcel post developed with a marked metropolitan bias.

As a result of this metropolitan bias, the ensuing integration refashioned rural areas into extensions of urban areas. According to Atwood (1986), "The rural routes 'citified the country' by diminishing rural cultural autonomy and by transferring the locus of public discourse and social interaction from small rural villages and county trading centers to larger towns and cities" (269). In effect, the rural areas were conscripted into the metropolitan system, and the very character of rural areas that RFD's proponents strived to preserve was lost. This prompted the editor of *Wellman Advance*, a local paper in Iowa, in 1901, to ask a profound question: "Why not keep our country post office and let the rural routes go?" (Atwood 1986, 269).

In the case of other systems, as we will see in other chapters of this book, the development of universal access for the postal system was marred by blind spots. They resulted from a lack of understanding of the double-edged nature of connectivity and inattention to the system's gains. The expansion of the postal service in rural areas was different on this score. Small businesses and local governments were aware of the downside of connectivity, which was very evident in their initial efforts to limit connectivity to businesses in the nearest town. However, they were unable to secure buy-in from rural consumers, drawn to lower mail-order prices and ill disposed toward local business for charging higher prices over the years.

But then the expansion of the postal service undermined what the proponents sought—preservation of rural communities. It enabled changes that transformed the character of rural communities. The problem here was that the proponents' preoccupation with the farmer's gains distorted their field of vision, and the possibility that urban interests could be much bigger winners was not given due consideration. The decentering of the farmer would have opened up a new field of vision and enabled a fuller understanding of the forces at play.

The curious complaint in 1909 of magazine publishers with regard to postal delivery to northern Alaska was a window into what eventually transpired. It deserved more than a passing reference in a long *New York Times* article extolling the gains of the isolated Alaskans. Similarly, in the policy discourse at large, the system's gains and the individual's travails deserved much greater attention.

3

EDUCATION

The Declaration of Independence famously proclaimed it to be a self-evident truth that "all men are created equal." Subsequently, after the birth of the US, its Constitution guaranteed its citizens equal rights such as right to free speech, religious freedom, and peaceful assembly—limited in its inclusivity and imperfectly put into effect by entrusted authorities . But, in practice, citizens' capacity to exercise their constitutional rights has always been unequal, given persisting economic inequalities. When lack of resources undermines someone's constitutional rights, fundamental questions arise about justice, democracy, and the constitutional order, creating a tension that animates American politics, especially issues concerning universal education (Nelson 1982; Thomas 1967; Weisskopf 1975).

In the introductory note to an article on universal education, *Mechanic's Free Press*, published in Philadelphia from 1828 to 1831, rudely shone the spotlight on the aforementioned contradiction:

We consider the present a subject of exceeding great importance: it involves a principle of natural right—should we be deprived of *that*, our Freedom is mockery, and our Constitution a dead letter.

It will be seen the subject is beginning to be freely discussed in our columns, and we hope it will receive the serious attention of the Legislature (1830, 3, italics in original).

Over a century later, in 1973, Justice Thurgood Marshall raised the same issue in a dissenting opinion in *San Antonio Independent School District v. Rodriguez*, where the Supreme Court upheld Texas's system of financing public schools in part because it did not consider education to be a fundamental right. Justice Marshall, taking education to be a necessary precondition for the meaningful exercise of the right to free speech, a constitutionally guaranteed fundamental right, asked the majority: "I would like to know where the Constitution guarantees the right to procreate . . . or the right to vote in state elections . . . or the right to an appeal from a criminal conviction" (Nelson 1982, 725).

Given this tension arising from persisting economic inequality, the notion of equality of opportunity is attractive, for it opens the possibility of finding a workable solution. It emphasizes *initial* equality of opportunity. It also justifies disparities in income on basis of merit if equality existed initially. In such a context, universal education has a special appeal, as it creates initial equality, or at least it appears[1] to do so.

If the Constitution had made education a constitutional right or said something specific about it, the story of universal education would have been very different. But that was not the case—it does not even mention education. For this reason, the proponents of universal education invoked provisions of the Constitution that are only indirectly related to education (Carlton 1906; Nelson 1982; Thomas 1967; Weisskopf 1975). The jousting on this issue has shaped the development of universal education in ways that are very interesting for our analytical project, as we will see in the discussion that follows.

EXPANSION OF EDUCATION

England's Poor Laws of 1563 and 1601 governed the early efforts of the American colonies to expand education.[2] They enunciated two important principles: the state is the authoritative actor for the development of education, and everyone should have a minimum level of education (Kotin and Aikman 1980). The actual path to the development of universal education was a long and complex one.

Massachusetts enacted a compulsory education law, the first such law in the colonies, in 1642, requiring all parents and masters to ensure that children in their care were taught a trade and reading. Town officials were to check and, in cases of noncompliance, move children from the custody of their guardians to responsible persons. We need to note two points here: One, provision of education was the responsibility of parents and masters. Two, education was to be provided to *all* children (Kotin and Aikman 1980).

In 1648, Massachusetts enacted an amendment to the law, which, among other things, introduced a new provision—it authorized the use of monies from the town treasury to pay for a master. This provision was critical for the development of universal education, as it established the principle that monies could be raised through local taxation to pay for schools and teachers. The other New England colonies, except Rhode Island, adopted variations of the Massachusetts model (Kotin and Aikman 1980).[3]

Virginia adopted a very different approach, directing its education laws specifically at poor and socially marginalized children—orphans, illegitimate children, mulattoes with white mothers, and indigent children. As for the rest of the population, it took education to be a private matter. Other southern states adopted variations of the Virginia model (Kotin and Aikman 1980).[4]

Over time, the colonies increased the minimum level of education specified in their universal education laws. A 1642 Massachusetts law specified that all children should be taught to read,[5] and in 1660, New Haven Colony added a writing requirement.[6] Subsequently, the colonies expanded the scope of universal education to history, algebra, geometry, bookkeeping, surveying, logic, Latin, Greek, rhetoric, and other subjects (Kotin and Aikman 1980; Meyer 1965). Such endeavors were local, with the colonies basically providing enabling legal frameworks, especially for taxation (Kotin and Aikman 1980; Sawhney and Jayakar 1999, 2007).

Within this general framework, which catered to local idiosyncrasies, several forms of schools proliferated. Schools providing primary education included dame schools (run by women in their homes), old field schools (located on exhausted land and run by teachers under contract), and parochial schools (run by churches). Venture schools provided training in trades, accounting, and other skills of intermediate professions, such as plumbing and carpentry. Grammar schools taught boys classical languages, preparing them for college and subsequent careers in law, medicine, clergy, and other high professions. Later, they also started teaching more practical subjects. Finishing schools taught upper-class girls etiquette and the social graces, preparing them for entry into society. Among other forms of schools, notable ones included town schools and endowed free schools (Good 1956; Hillesheim and Merrill 1971; Webb 2006). Further, schools were funded by "nearly every money-raising scheme known to man"—tuition, lotteries, fines for drunk and disorderly behavior, sales of war booty, and license fees, among others (Madsen 1974, 88). Adding to the institutional diversity, communities borrowed ideas from each other, often modifying them in innovative ways (Sawhney and Jayakar 1999).

A few decades after such elements coalesced into an elemental system of sorts in the mid-seventeenth century, a decline started. Many colonies modified and subsequently repealed laws requiring compulsory education, and in colonies where they remained in place, their enforcement weakened (Kotin and Aikman 1980).[7] After the US gained its independence, the hopes and ambitions of the new nation fueled a revival of the discourse on universal education, seen as critical for a successful democracy.

Each of the first six American presidents, from George Washington to John Quincy Adams, called for universal education (Counts 1967). But they were not able to do much about it because the Constitution had essentially left education to the states. Thomas Jefferson, a strict constitutional constructivist, sought a constitutional amendment to allow a federal role in education, but that did not come to pass. Over the next century or so, support for federal engagement mostly declined. It was only in the New Deal era that the present-day federal education policy and administration apparatus started taking shape (Thomas 1967). Its legitimacy remains a nagging question and a persisting aggravation in strict constitutional constructivist circles.

INEQUITIES AND INEQUITIES THAT MATTER

The founders of the new nation saw universal education from the perspective of a democratic polity or, in our analytical terms, a system-centric perspective: important for its flourishing, "primarily in *political* terms rather than in terms of . . . individual fulfillment" (Butts 1978, 364–365, italics in original). For them, universal education was critical on two accounts:

(1) it was essential for informed participation of voters in the democratic process,[8] and (2) it would help bind the new country together.[9] In this scheme of things, in the words of Benjamin Rush (1965), education was a means to "convert men into republican machines" so that they could fit in and serve the "great machine" of the republic (17).[10]

Despite this strong sentiment, nothing significant about universal education happened on the ground. This was at heart a failure to "translate sentiments into appropriations" (Ditzion 1947, 10). The mismatch between words and commitment of necessary resources was particularly dramatic in Indiana. Indiana's constitution, adopted in 1816, directed lawmakers to set up an educational system that is "gratis and open to all." However, it also permitted them to defer the endeavor until "circumstances will permit" (Constitution of the State of Indiana, 1816, Article IX, Section 2). This qualifying clause accommodated inaction, and lawmakers put off the actualization of these ideals for many years.

But then, in the mid-nineteenth century, universal education started mattering to states and local communities, motivating them to deploy the necessary resources for it. Their reasons for making a concerted effort are important for our analysis.

The industrialists needed trained workers. The labor leaders were eager to remove children from the workforce, as their availability was depressing wages, during a time of large influxes of immigrant labor.[11] Apprehensive of "mob rule," the propertied elites saw universal education as a means of "reconciling freedom and order"—that is, indoctrinating the newly enfranchised lower classes to buy into the established system (Kaestle 1983, 5). Ordinary people, motivated by their fear of Catholic immigration, were willing to pay taxes, seeing the tax-supported public school as the "principal digestive organ of the body politic" for Americanizing Catholics and countering their proclivity to set up parochial schools (Strong 1963, 89). Moreover, immigrants, even Protestants, needed to learn English to be effective workers. In effect, universal education was a product of the convergence of multiple agendas (Binder 1974; Button and Provenzo 1989; Gabriel 1964; Sawhney 1994).

Such a large coalition was needed to get the ball rolling because the opposition to universal education as public policy was fierce. Orestes Brownson, the editor of *Boston Quarterly Review*, articulated an early critique on intellectual grounds. He was not against education for everyone per se: "The real question for us to ask is not, Shall our children be educated? but, To what end shall they be educated, and by what means? What is the kind of education needed, and how shall it be furnished?" (Brownson 1839, 394). He saw in universal education an effort to build a system that molded children in particular ways—well-adapted citizens and productive workers. He believed that if this were to come to pass, we would "lose the man in the citizen" (Lasch 1991, 189). He argued that the primary focus of education should be "the man," and such an education could flourish only in locally run, autonomous schools. In contrast, universal education as public policy would lead to centralization of administration, standardization of curriculum, and professionalization of education—all of which would be intolerant of idiosyncrasies of persons and locales that produce education in nonparticular ways.

On the ground, proponents' universal education initiatives met with fierce political opposition. For instance, in Massachusetts, Democrats saw universal education as a Whig project, having taken root under Whig governors and Whig-dominated legislatures. When the Democrat Marcus Morton won the governorship in 1839, he tried to dissolve the Massachusetts Board of Education, established under his Whig predecessor's leadership. In his address to the legislature, he emphasized the need to retain local control in the "town and district meetings, those little pure democracies" (Kaestle and Vinovskis 1980, 215). He and his party were opposed to state-level centralization and professionalization of education. However, the bill to abolish the board failed by a 182–245 vote. About one-third of the Democratic legislators voted against the bill, saving the board; on the other hand, about one-fifth of the Whig legislators voted for it—these crossover votes are notable, as they give us a measure of the contentious complexities of universal access. Thenceforward, advocacy for universal education continued (Kaestle and Vinovskis 1980; Taylor 2010).

This push for universal education reactivated the old debate, but with much greater energy, as it was no longer about a distant ideal. It also raised new issues.

MOVEMENT FOR UNIVERSAL EDUCATION

The philosophical tension between approaches that sought to cover all children or focused only on marginalized ones, first seen in the contrasting Massachusetts and Virginia models, was still alive and well.

The adherents of the expansive approach argued that universal education was in essence an investment in the flourishing of the society as a whole—enhancing both the democratic process and economic productivity (Krug 1966). In 1835, *American Monthly Magazine* opined: "Viewed in a political light, education is imparted, not for the sake of the recipients, but of the state of which they are members" (Jackson 1941, 63).[12] Further, education had to be universal—cover all children—not only because it was morally right and economically beneficial, but also because it allowed for a system, wherein monies could be moved in ways that optimized the overall well-being of the society (Binder 1974).

The opponents of the expansive view argued that education was a private matter (Meyer 1965). Parents should educate their children as they see fit within their particular circumstances—the state had no right to impose its dictate. Furthermore, they saw universal education as a wealth redistribution scheme, property tax–funded schools as an infringement on the right to own property, and forcing people to pay for other's children as morally questionable (Binder 1974; Carlton 1906). In their view, if state intervention in education were to occur, it should be limited to the education of marginalized children.

Reformers urged critics to take a longer view beyond "others' children" and shoulder some responsibility for the welfare of future generations, as they needed to repay their own debt to preceding generations (Binder 1974).[13] The preeminent reformer Horace Mann, on appointment as the first Secretary of Massachusetts Board of Education, wrote in his private journal: "Henceforth, so long as I hold this office, I devote myself to the supremist welfare of mankind upon earth" (Kaestle and Vinovskis 1980, 209). Mann had to reach beyond this own humane understanding of universal education to other logics that would appeal to different groups in order to make it a reality on the ground.

Appealing to the more direct self-interest of the elites, Mann added that educated multitudes were less likely to be drawn to radical promises of the opponents of the established order (Binder 1974; Button and Provenzo 1989). Other commentators built up this basic argument in the following manner. One asserted: "Education—free, universal education" was the only security for one's possessions "in a republic." Another directly asked "whether the schoolmaster be not as essential a protector of life and property as the constable or the policeman." Leveraging fears of bodily harm, another commentator characterized education as an "antidote" against unwarranted "hatred by the poor toward the rich" (Jackson 1941, 67 and 68). This line of argument resonated with the elites, who were unsettled by the extension of suffrage in the Jacksonian era to all white males whether or not they were propertied men, and fearful of untutored masses upsetting the established order, especially on private property–related issues (Jackson 1941). For our analytical purposes, it is telling that Horace Mann characterized education, a supposedly liberating endeavor, as "the balance-wheel of the social machinery" (quoted in Binder 1974, 50).

In this context of overlapping interests, we start to see the emergence and popularization of the common school—tuition free, tax-funded, nondenominational, and universal. It required legislation against child labor, on the one hand, and for compulsory attendance, on the other. It also required the passage of laws for levying taxes, for which the school district system was set up. Correspondingly, for the implementation of these mandates, state and local governments created educational bureaucracies. As a result of such state and local efforts, with very little involvement by the federal government, great advances were made between 1825 and 1850 in universal education. By the mid-nineteenth century, it was an established principle (Thomas 1967).

GAINS AND TRAVAILS

The development of universal education was marked by complex and contentious trade-offs. On the one hand, we have the very nature of education being child-centric or system-centric—the former cultivating the idiosyncratic flourishing of each child, and the latter producing a "particular kind of citizen" (Taylor 2010, x). In a healthy democracy, while we might expect a workable tension between the two endeavors, the reality is complex and conflicted (Taylor 2010). On the other hand, we have the trade-off on who should control a child's upbringing: the parents or the state. With compulsory attendance laws, parents essentially had to cede control over their children, as the state was defining what constitutes "education." Lacking resources for an authorized alternative source of education, parents typically send their children to tax-supported public schools. It is telling that a large number of parents choose to homeschool their children with their own resources—even here, however, they have to satisfy state laws and regulations. Given such trade-offs and the obfuscations that come with them, in the case of education, our exercise of taking stock of the gains and travails for the individual and the system has special value (table 3.1).

INDIVIDUAL

Gains The individual benefited from increased earning power, greater awareness of the world, and skills to function in the world beyond the immediate social environment. In addition to economic benefits, the individual

Table 3.1 Universal education: Gains and travails

	Gains	Travails
Individual	Increased earning power Skills to function in the world beyond one's immediate environment Social benefits such as improved sanitary practices, family planning, and women's empowerment	State's encroachment on parental control of children Secularized education that does not offend anybody Centralization and bureaucratization of education Prioritization of the needs of state and economy over the idiosyncratic flourishing of children
System	Educated workforce Increased capacity to create and staff complex systems in industry, military, medicine, etc. Savings such as in health care, with improved personal hygiene Assimilation of immigrants	Unavoidable trade-offs of the sort that give rise to unhappy constituencies Expansionary tendency—there are always constituencies pushing for upping the level of universal education, to higher grades in school, to community college, to four-year college

gained from social benefits such as improved sanitary practices, family planning, and women's empowerment. We need not belabor the gains from universal education—what we need to note for our purposes is that in the aggregate, it improved the material conditions and the quality of life of the individual.

Travails   Today, the debate is limited to the type of universal education, not its very existence. In effect, the scope of our thinking has narrowed, as universal education is now a given in our considerations. To step out of the confines of our present-day understanding, we now revisit the losing arguments of the opponents of universal education.

Looking back at the intensity of opposition to universal education, Thomas (1967) characterizes it as "second in intensity only to the struggle to abolish slavery" (14). The very notion of universal education vexed its opponents, especially on the following four counts.

One, universal education entailed state encroachment on parents' control over their children's upbringing. In that view, education has always been a private matter and should remain so on moral grounds.[14]

Two, universalization necessitated a type of education that did not offend anybody. Opponents saw therein an in-built tendency toward secularization; in the words of Orestes Brownson (1839): "Christianity ending into nothingness" (404). Horace Mann insisted that that was not his intent, and his actions suggest that he indeed intended to operate within the framework of Christianity. But that limited notion of universality came apart with the arrival of immigrants of different religious persuasions. On the long arc of history, the opponents were proved correct.

Three, universalization necessitated large-scale coordination. Opponents saw in this an in-built tendency toward centralization. Mann strongly argued for systemization of the variegated mix of public and private schools to eliminate redundancies and bring down the costs. He also argued for cross-subsidies from rich schools to poor schools to maximize the overall social benefit (Taylor 2010). As the universalization project moved forward, systemization moved from the local to the state level. Eventually, it reached the federal level, albeit to a lesser degree. At each stage, the impetus for increased systemization was need for increased efficiencies and cross-subsidies.[15]

Four, the opponents argued that universalization was prompted by the needs of the body politic and economy, not the welfare of children. By the mid-nineteenth century, the emphasis had moved from the bonds of nationhood to the citizen's ability to function effectively in a republic. In Mann's view, "It may be an easy thing to make a Republic; but is a very laborious thing to make Republicans" (Taylor 2010, 29). Mann and his followers felt that the founding of the republic was an unfinished project, as the unprecedented constitutionally guaranteed freedoms needed to be matched by the cultivation of citizenship. Otherwise, left alone to human nature, these freedoms were likely to elicit more base than noble behaviors.[16] The framers of the Constitution had sought to counter this tendency with a system of checks and balances. But in real life, these checks and balances usually took the form of one base behavior being countered by another. Universal education could raise the level of interaction within the body politic (Taylor 2010).[17]

In effect, "free" universal education has its travails: loss of the religious ethos (major distress for the religiously minded), loss of diversity of institutional forms (major loss for the locally minded), and the imbalance between the needs of society and the idiosyncratic flourishing of a child (major deprivation to the child, and also to adults who prize realizing the true potential of each child). The strenuous efforts of homeschoolers, who take great pains to keep their children out of mainstream school systems, remind us that free universal education imposes significant travails.[18]

Mann himself characterized children as "ductile and mouldable" "materials" upon which education "operates" (Taylor 2010, 30). Notably, similar logic also applied to the production of factory workers, a major factor behind the push for universal education.[19]

SYSTEM

Gains Universal education produced an educated workforce for industry, which as a whole was more valuable than the sum of its parts (i.e., individual workers). Francis Wayland, an economist, in 1837 argued that "improvement in knowledge, in order to be in any degree beneficial, must be universal. A single individual can derive but little advantage from his

knowledge and industry if he be surrounded by a community both igno-
rant and indolent. In just so far as other men improve their condition,
and become useful to themselves, they become useful to him; and both
parties thus become useful to each other" (Carlton 1906, 55). Wayland
was focusing on the gains that individuals derive from higher-quality
interactions among themselves, but the system is also a major, if not bigger, beneficiary.[20]

Education enables the complex managerial and peer interactions necessary for the running of complex systems in industry, military, medicine,
and other realms of an industrial society. It also helps contain the negative
fallout of interactions among individuals (e.g., improved personal hygiene
reduces contagions, and thereby public health costs). Finally, it works to
bring everybody—especially immigrants—into the fold, enabling the most
optimal utilization of "human resources."

Travails Universal education is a complex undertaking requiring many
trade-offs, which invariably upset some group of people. Furthermore,
some constituency or other is perpetually pressuring for upping the level of
universal education. Consequently, over time, universal education inched
up to the high school level in the early twentieth century. That seemed
to be a natural plateau until World War II, when US had to scale up its
army expeditiously. In the course of this buildup, the military found that
it did not have an adequate pool of potential recruits for its officer core.
To work around this bottleneck, it employed aptitude tests to identify
recruits with officer potential and gave them accelerated training (Bowles
1966).

While this program served that immediate national need, its success
also created disquiet among policymakers because it showed that the US
was wasting talent.[21] This realization spurred policymakers to expand
access to college for veterans with G.I. Bill monies and other initiatives for
disadvantaged students.[22,23] More recently, in 2014, Tennessee launched a
program offering tuition-free education at a community college or a technical school for two years.[24] In 2015, Oregon launched its own program
(Manning 2015). Subsequently, many other states have followed suit,
including Arkansas, Delaware, Kentucky, Minnesota, Montana, Rhode
Island, Nevada, and New York (Powell 2018). Now, there is even talk of
offering tuition-free access to four-year colleges.

DECENTERING THE SYSTEM

In the case of universal education, the key questions have been: Who gains more—the individual or society? Who should pay for it? Herein, the individual's travails have mostly been overlooked.

Kaestle (1983), writing of the development of common schools from 1780 to 1860, nicely tees up the issue we need to consider: "The morality of the social system as a system was beyond question; the moral quality of the society was therefore to be improved by improving the moral quality of individuals" (81). In other words, the nature of the social system was a given. It was so nonnegotiable that it was not even worth discussing. And we do hear a deafening silence on the nature of society in the universal education debate.

The focus is instead on "improving"—a euphemism—the individual. What we are really talking about is changing the nature of the individual to secure tighter alignment with the system. This so-called improvement is presented as a gain. The discourse pretty much works with the triad, discussed in chapter 1, of inclusiveness, access, and gain. This is not incorrect; it is incomplete. Another logic is also at play—that of conscription, conversion, and travail.

When we strip away euphemisms, we can see conscription at work. Consider compulsory attendance laws, truant officers, and other mechanisms that ensure school attendance by punishing the parents if their children do not attend. Of course, they are presented as being in the best interest of the children—safeguards for their well-being. This is not incorrect; it is incomplete. How about the other way around (i.e., they are in the best interest of the system)? Here, we do not have to read between the lines. We have a number of very explicit statements on this score, some of which have been quoted earlier in this chapter. One of these, by Samuel Harrison Smith, cowinner of the 1797 American Philosophical Society prize for best essay on a national system of education, fits well into the weave of the present discussion: "Society must establish the right to educate, and acknowledge the duty of having educated, all children. A circumstance so momentously important must not be left to the negligence of individuals" (Smith 1965, 190). What constitutes "the negligence of individuals" from a system-centric perspective could be "self-determination" from an individual-centric perspective. Parents may genuinely believe that it is not in the best interest of their children to attend a school of the type provided by the system. This is not an idle theoretical point, as nearly two million parents in the US today choose to homeschool their children instead of sending them to public school.

Conscription is followed by the conversion of children into a resource for the system. The system's imperatives are particularly vivid in the following argument, among others, advanced by Horace Mann for universal education: "Education must prepare our citizens to become municipal officers, intelligent jurors, honest witnesses, legislators, or competent judges of legislation,—in fine, to fill all the manifold relations of life. For this end, it must be universal" (Mann 1964, 98).[25] Indeed, the education system prepares youth for these roles, teaching them reading, writing, mathematics, science, civics, and other important subjects. However, more fundamentally, as Foucault (1995) penetratingly informs, it teaches them to live by the schedule and the bell. This education is so deep that even after what they were taught (e.g., the periodic table) fades in their memories, the students never forget the schedule and the bell—it becomes second nature to live by them, the overlays that align them with system modalities.

The individual's travails in all this are rarely ever talked about. What about the personal flourishing of the young? Human flourishing requires space for idiosyncrasies that make us human in our particular ways. But idiosyncrasies are the bane of systems. The system-centric logic is particularly visible in "A System of Equal Education," by Robert Coram, a Revolutionary War veteran. As Messerli (1967, 422) relates, in 1791, "Coram proposed that the country be sectioned into six-mile squares with each being served by a school. He made no adjustments for existing schools nor concessions for local preferences or customs. As a matter of fact, the latter were seen more as a threat than an asset." This is an extreme case, but shades of such thinking inform all visions and plans for universal education; such is the nature of systems and their imperatives.

Coram's system regimen is rudely visible against the materiality of a spatial expanse, something that is easily visualizable. System regimens also work on the time dimension, being woven into everyday life such that they are difficult to discern. The mores of punctuality are such regimens. They make modern humans behave in a punctual manner that is artificial so as to align them with the imperatives of the system. This affliction is a product of systematic application of disciplines (education, preaching, fines, etc.), starting with industrialization. Save for the last couple of centuries, humans were not afflicted with compulsive punctuality. In the nonindustrialized parts of the world, even today humans are not punctual. But for us in the industrialized world, punctuality has become second nature (Foucault 1995; Thompson 1967; Zuboff 1988)—so much so that we will abruptly curtail a joyful conversation, which rarely occasions us, at an appointed hour. Punctuality is good for the operation of systems, but is it good for human flourishing?

This critique does not do justice to the totality of universal education. But it should not be ignored. The fact that the universal education movement was to a noninsignificant degree driven by the fear of newly enfranchised untutored masses upsetting the established order gives credence to it. The other major motivation was need for an educated industrial workforce, which also begs for a critical view. This is not to deny noble motives, but one needs to be mindful of the mix of motivations that rally humans.

Now, if we decenter the "negligence of individuals" that needs to be overridden with a systemic solution, and center-stage children and their flourishing, a different view emerges. We find ourselves aligned with Thomas's (1967) critique of Horace Mann: "His emphasis, as we have seen, is generally less on children's right to an education than on the political community's need for educated citizens" (34). Herein comes a critical question: should children be molded to enhance the well-being of society, or should society change to enhance the flourishing of children? Much thought and effort have been devoted to the former part of the question, but the latter also deserves attention, as it would help us find a happy middle ground. In that case, "concessions for local preferences or customs" should be made, the regimens loosened, and the scale-up of political and productive processes across geographical expanse reduced—even if it means less riches.

4

ELECTRIFICATION

"The best lighted and heated city in the country." This was the claim of an 1892 pamphlet promoting Muncie, Indiana—a city of such a typical character that social scientists referred to it by the generic name "Middletown." In this manner, the electrification of Muncie and its hinterland, as the historian David Nye (1990) describes, also embodies the story of electrification across America.

On March 31, 1880, with the lighting of four arc lights on the courthouse, Wabash, in north central Indiana, said it was the "first electrically lighted city"—a claim that an Indiana Historical Bureau site marker carries to this day. This event, witnessed by 10,000 visitors and correspondents of over forty newspapers, brought the world's attention to Wabash. The citizens of Muncie, piqued about being overshadowed by a smaller Indiana town, installed 100 arc lights in 1892. In 1894, with 132 such lights in town, *Muncie Morning News* crowed that "there is not a better city electric lighting plant in the state" (Nye 1990, 6). In 1885, James Boyce, a wealthy businessman, illuminated shops with electric lights in his downtown buildings, attracting crowds (Kemper 1908). In 1888, the city council granted its first license for a trolley system, which grew into an interurban network connecting nearby communities, with Muncie in the center (Nye 1990).

Clearly, Muncie accomplished much within a decade, but the process was fitful, with many ups and downs. To start with, since gas was readily and cheaply available in Indiana, most enterprises preferred the continued use of gas, switching to electricity only after the gas fields petered out at the turn of the century (1901–1907). In the case of residential use, the uptake was much slower—only twenty-two homes had electricity in 1899. However, by 1926, 95 percent of homes had electricity. Muncie's metamorphosis into a sparkling urban center sharpened the contrast with its still dark rural hinterland[1] (Nye 1990), literally and also metaphorically—with electricity representing progress (Lieberman 2017; Marvin 1988). Despite the contrast, the Indiana General Service, deeming rural service to be uneconomical, was unresponsive to requests for it. It would offer service only if rural customers guaranteed five years of usage, amounting to 18 percent of the build-out cost. Few rural customers could afford to do so. Consequently, by 1931, of all the utility's customers, only 4 percent were in rural areas (Nye 1990).

In effect, after electrifying streets, commercial establishments, trolley systems, factories, and residences in urban areas, the utilities were reluctant to undertake the last stage—rural electrification—which was essential for universal access. This shortfall did not sit well with President Franklin D. Roosevelt's administration, creating tension with the utilities. In effect, the development of electrification was marked by the conflicting logics of the utilities and the proponents of rural electrification, as well as by blind spots in their logics, as we will see in this chapter.

EXPANSION OF ELECTRIFICATION

Electrification started with hyperlocal endeavors, such as factories installing dedicated generators for their own use. In 1882, Edison Illuminating Company established the world's first central power plant, Pearl Street Station in New York, with the goal of supplying electricity in a service area of one square mile (Hughes 1983; Rudolph and Ridley 1986).[2] Upon commencing service, Edison offered it for free, seeking to understand the distribution and use of electricity before entering into contracts with customers.[3] A month later, he had 59 "customers;" three months later, he began charging, and a year later he had 513 customers (Hughes 1983; Munson 1985; New York Edison Company 1913). Thereafter, Edison and others started establishing central power stations in other parts of the country. The next proof of concept for centralized power generation was the Niagara project, which in 1895 demonstrated the viability of generating power at scale in one place and then distributing it over long-distance transmission lines (Meinig 2004). Subsequently, in an incremental, bottom-up, and contested process, power generation and distribution companies connected their plants and distributions lines into regional grids (Bakke 2016; Cohn 2017).

The expansion of electrification was marked by what Hughes calls reverse salient and momentum.[4] Hughes borrowed the term "reverse salient" from military historians, who use it to refer to parts of an advancing army that fail to keep up, attracting the attention of the top brass, who then direct resources to beef them up. Hughes (1983) found the metaphor "appropriate because an advancing military front exhibits many of the irregularities and unpredictable qualities of an evolving technological system" (14). In the case of technological systems, reverse salients arise because of the imbalances in the capacities of various system components. The resulting suboptimality attracts institutional attention and resources for technological breakthroughs that advance the capabilities of the lagging components. The removal of a reverse salient allows further system growth, which in turn increases complexity and gives rise to a new reverse salient.

While reverse salients are markers of system growth, "momentum" refers to direction and pace of the growth. As a technological system grows, institutions and individuals become invested in its growth: corporate and noncorporate owners and operators of the system, regulatory agencies overseeing them, education establishments producing skilled workers for them, and workers trained in accordance with system needs. Drawing on the experience with polyphase electric systems, Hughes (1983) observes that "the systematic interaction of men, ideas, and institutions, both technical and nontechnical, led to the development of a supersystem—a sociotechnical one–with mass movement and direction. An apt metaphor for this movement is 'momentum'" (140). The bigger point is that once interests are aligned in a certain way, it is difficult to change the direction of system growth.

In this fashion, Hughes provides a detailed account of reverse salients and systemic momentum as the system grew from Edison's Pearl Street

Station to regional grids. These details are not important for our analytic exercise here. We need to note the nature of systemic expansion: between 1915 and 1932, system expansion broadened residential service from 20 to 70 percent of Americans nationally, but only 10 percent lived in rural areas (Meinig 2004). From this point on, system expansion was a product of public policy intervention, *not* momentum.

INEQUITIES AND INEQUITIES THAT MATTER

In the utilities' view, the cost of constructing rural lines was prohibitive. Their thinking centered on the notion that each line should pay for itself (Reed 1935; Stauter 1973). On the demand side, in their estimation only dairy, irrigation, and a few energy-intensive types of farming could be served profitably (Meinig 2004).[5] In this view, the conundrum was as follows:

Cost was the real stumbling block to service. Rural lines cost $2,000 or more per mile, and since there were usually only two to five dwellings per mile in the country, utilities anticipated low revenue to amortize investments. . . . Companies expected farmers, therefore, to bear the burden of the initial investment charging them with the cost of the line, or a $500 to $1,000 deposit. Rural rates were also high, about 9 to 10 cents per kilowatt-hour for the minimum usage. . . . Few rural homes could afford to pay for the lines or make the deposit, nor could they at first afford enough appliances to use the amount of electricity necessary to achieve the advantage of lower rates. The effect was an endless cycle of expense for both parties—recipients of service used little power because of high rates, and the utilities charged such rates because of low usage. (Brown 1980, 5)

In effect, rural electrification was taken to be an irresolvable chicken-and-egg problem: reduction in costs would require increase in demand, which would generate economies of scale, while increase in demand would require reduction in costs, which would make the service affordable.

In sum, while the utilities accepted that provision of electricity in rural areas would greatly enhance the quality of life, they could not see how it could be an economically viable proposition. In their thinking, "electric distribution was what it had to be and little could be done about it" (*Electrical World* 1935, 56).

To those invested in the human side of things, this was an inequity that mattered so much that it trumped pure economic considerations. They found the drudgery of rural life unacceptable when labor-saving technology was available (Childs 1974; McCraw 1971). A statement from Murray Lincoln, secretary of the Ohio Farm Bureau, was particularly compelling: "Why should we sentence men and women to do by the sweat of their brows what electricity can do so much easier and cheaper? After all, human beings ought to be of more importance than returns on utility stocks" (*Rural Electrification News* 1936b, 28). The National Electric Light Association estimated in 1912 that every year, on average, farm wives were spending twenty more days washing clothes than their urban peers with electric washers. Further, they had to use sad irons[6] to press the clothes (Brown 1980). In this fashion, the sad iron became an evocative metaphor and a rallying cry, as "sad" comes from "sæd" (meaning "sated, weary" and also "weighty, dense")[7] in Old English, and also carried connotations that resonated with their cause.

By the 1930s, the notion that electric service is a basic human need had struck deep roots. In 1932, in a campaign speech for the presidency, Franklin D. Roosevelt, then the governor of New York, declared that "electricity is no longer a luxury—it is a definite necessity . . . It can indeed relieve the drudgery of the housewife and lift the great burden off the shoulders of the hardworking farmer" (Roosevelt 1932, 13).

Moved by quality-of-life considerations, the champions of rural electrification directed their energy at finding ways to cut through the chicken-and-egg problem, as opposed to walking away from it. On the economics plane, their goal was not profit per se, but rather an economically sustainable model that eased the burdens of rural life. Their determination and talents brought about much-needed breakthroughs—conceptual and technical. Morris Cooke, the great champion of rural electrification,[8] believed that there was always a "practical way to get done what needs to be done" (Childs 1974, 47).

In this way, champions of rural electrification started focusing on distribution. They argued that the utilities focused on generation and transmission—parts of the network of greatest economic import—and neglected distribution, which was of particular significance for rural electrification. As a starting point, they examined the economics of distribution, where they found the available data to be illegible from the standpoint of cost accounting (*Rural Electrification News* 1935a). Moreover, they noticed significant disparities in cost accounting practices across companies, which suggested that something was fundamentally wrong. In particular, the existing cost accounting systems no longer seemed valid, as technological advancements had changed the cost parameters and patterns of demand (Childs 1974).

PROPOSAL FOR AREA COVERAGE

The proponents saw two fundamental flaws in experts' conception of rural electrification: (1) the notion that it entailed the extension of long lines from urban systems into rural areas, and (2) the assumption that the rural households would have only limited use for the service extended to them.[9] In response to the former argument, they said that rural electricity should be seen as a system on its own, not as a mere appendage of the urban system. They called for "area coverage" (the electrification of swaths of land in an integrated manner, as opposed to the extension of isolated lines from urban systems), arguing that the resulting economies of scale would bring down the costs dramatically below the utilities' per-line calculations (Burritt 1931; Stauter 1973). As for the second concern, they scoffed at the notion that rural households would use only a sixty-watt light bulb's worth of electricity, pointing out that as places of both residence and work, they were likely to have even greater usage than urban locales (Stauter 1973). Furthermore, they noted that after securing electricity, rural households tended to purchase appliances, which significantly increased their electricity usage (Brown 1980).

This new thinking gave rise to the Rural Electrification Administration (REA), established by President Roosevelt in 1935 under the authority of the Emergency Relief Appropriation Act of 1935.[10] Upon its establishment, REA sought to develop a partnership with electric utilities to implement its program. It was only when these efforts failed that it turned to rural cooperatives (Christie 1983).

REA channeled the American self-help ethos, in marked contrast to other countries where governments themselves constructed rural networks. This was remarkable, given that only 11 percent of American farms were electrified in 1935 and the corresponding figures were much higher in other countries: 95 percent in the Netherlands, 90 percent in France and Germany, and 80 percent in Denmark (Coyle 1936; Nye

1990). Many groups called for direct government intervention to catch up with other industrialized countries (Childs 1974; Slattery 1940). Yet the REA embarked on a bottom-up strategy.

REA provided subsidized loans—with softer interest rates,[11] payment periods, and collateral requirements—to rural cooperatives for building, maintaining, and operating their own distribution systems. With these self-liquidating loans, REA worked to establish eventually self-sustaining electricity cooperatives—a venture in social entrepreneurship that its third administrator, Harry Slattery, characterized as a "business like any private enterprise" (Slattery 1940, 6). In effect, REA saw itself as an incubator of projects rather than a source for permanent subsidy. In addition to loans, REA provided technical guidance to the cooperatives. As Cooke (1935) explained: "It is always well to remember that the REA can only give a *start* to the vast undertaking of electrifying rural America along progressive lines. But if it can give impetus to new policies and sound technique, I am confident the movement will go forward on its own momentum" (4, italics in original).

REA's area coverage approach generated economies of scale, which brought down the construction costs. Moreover, Cooke directed that "rural lines need not be built for the ages" (Christie 1983, 174). Accordingly, REA developed steel-reinforced poles without cross arms to lower maintenance costs, doubled the cables' span length to reduce the number of poles, and designed sturdy meters and other devices for rural conditions (McCrary 1939; Rural Electrification Administration 1938).

For their part, farmers contributed their labor, and their cooperation brought down the cost of rights-of-way (Burritt 1931) and many other things. For example:

Some time ago a few of the systems in the sparsely settled areas found that the cost of meter reading and billing was one of the large items in the final cost of electricity. Someone suggested that each member read his own meter and mail the reading to the office. This was tried with success and reduced the reading and billing costs from an average of 25 cents to as low as 5 cents. (*Rural Electrification News* 1939a, 21)

With such innovations and community inputs, REA brought down the construction costs from $1,500–1,800 per mile to about $900 per mile (Christie 1983).

In addition to building networks, REA worked to "build up the psychology of generous use of electricity" (Morris Cooke, quoted in Christie 1983, 177). To do so, it focused on lowering rates, believing that this would increase usage and foster innovations (Christie 1983; Stauter 1973).[12] To enable individual access and build up demand, REA also provided loans to rural households for inside wiring and appliances (*Rural Electrification News*. 1935b). Furthermore, it helped manufacturers develop simple and less expensive appliances for rural markets (Christie 1983).[13]

Area coverage also forestalled capacity suboptimization, which tends to plague the development of rural electricity and other networks. Often, initial estimates underestimate demand, and, consequently, the built network turns out to be inadequate in the face of actual demand. As Smith (1931) notes, "Experience mindfully reminds us, it should always be kept in mind that the developments may surpass the ultimate assumed" (816). By plunging straight into the development of the overall system, REA's area coverage approach prevented this problem.[14] Subsequently, the rural cooperatives went on to prove the utilities wrong on both counts—construction cost and demand. But that is another story. Here, we need to focus on the gains and travails of rural electrification—both expected and realized.

GAINS AND TRAVAILS

Both the proponents and the utilities saw gains from rural electrification. The disagreement was primarily on the economics, with the utilities emphasizing the costs. As we take stock of the gains and travails (see table 4.1), we see that the reality was much more complex.

INDIVIDUAL

Gains   Rural electrification improved the quality of life of country dwellers—at least as we moderns understand it. They gained two to four waking hours and creature comforts such as refrigeration. Also, local industrial activity increased, at least initially (Brown 1970). Moreover, they developed innovative uses of electricity.[15] For instance, N. G. Norris, the owner of a large frog farm, facing high feed costs because frogs do not eat anything dead, found an innovative way of cutting his expenses.

Table 4.1 Rural electrification: Gains and travails

	Gains	Travails
Individual	Improved quality of life Increased productivity on farms	Increased productivity on farms, reducing the need for labor, prompting more migration to urban areas
System	Utilities supplying wholesale power to rural electricity cooperatives Electrical appliances manufactures profiting from the expansion of markets for their products	Tumult generated by strenuous and sustained political pushback by those opposed to government's involvement in electrification

In Norris's telling: "We put unusually bright electric lights at various places close to the surface of the ponds. These lights attract hundreds of thousands of insects at night. The frogs gather beneath the lights, stand on their hind legs and eat their fill" (*Rural Electrification News* 1936a, 25). Another farmer used an electric washing machine to shell peas (*Rural Electrification News* 1939b). More broadly, electrification helped dairies, vegetable and fruit processing plants, canneries, grinding mills, cotton gins, grain elevators, sawmills and lumberyards, nurseries, stockyards and slaughterhouses, game and fur farms, planing mills, and machine shops (Slattery 1940). By the 1950s, the extensive use of electricity in rural areas belied the apprehensions of the utilities.

After a sweeping overview of the facets of rural life touched by electricity, we need to gain a sense of the depth of change as well. To do so, for the purposes of this discussion, we will delve into one change at some depth—running water made possible by electrification. In 1919, the US Department of Agriculture (USDA) reported that rural families spent over ten hours per week pumping water and carrying it to their kitchens (Brown 1980). Running water not only eliminated such laborious chores, it also greatly improved sanitation and public health (Cooper 1940). According to one report, after the availability of running water, a rural school saw a 350 percent increase in handwashing soap use (Radder 1939).

Travails While the proponents succeeded with their average coverage approach, multiple-purpose thinking, and building up of the "psychology of the generous use of electricity" (assisted by low rates and loans for appliances), they had their blind spots (Christie 1983). In particular, they expected electrification to improve the quality of rural life on the one hand, and strengthen the rural economy on the other, and thereby stanch the loss of rural population[16] and even generate a back-to-the-farm movement (Carmody 1939; Deutsch 1944; Erdman 1930; Lilienthal 1939; Slattery 1940; Stauter 1973).

The hope that electricity would decentralize industry was not limited to rural communities. Many people across the world, including in urban areas, harbored this hope. For instance, in France, Alglave and Boulard (1884) felt that in the preelectricity era, societies concentrated industry and bore the travails of densely packed and polluted cities because the economics of small, coal-fired engines were prohibitive. "Electricity, on the other hand, does not suffer the same losses in being divided so as to be put at the disposition of the humblest" (Alglave and Boulard 1884, vi–vii). Therefore, with this new source of energy, the disadvantage of small producers vis-à-vis big producers would be greatly mitigated, allowing for a new flourishing of small producers. In rural America, this hope was expressed as follows:

The effect of good roads and automobiles has been to centralize many kinds of industry in the towns surrounding the great cities. Goods that used to be made on the farm are now made in town and bought for cash by the farmer. With electric equipment of his own, the farmer can bring back some of these profitable activities to his own house and barn, saving himself trips to towns as well as money. With his own feed-grinding mill, his own refrigerator, his own fertilizer mixer, the farmer can often process his own materials for his own use, saving transportation and expense. The farmer's wife, with suitable electric equipment, will find that canning and preserving can be done with far less labor and discomfort than under the old methods. (Cole 1936, 16)

With such value-adding activities on the farm, the farmer was expected to move up the value chain and earn more on the one hand, and reduce damage and wastage of the produce in storage and en route to the market on the other (Childs 1974). Looking beyond agriculture,[17] the proponents hoped to entice industries to relocate to rural areas, with their lower rents and labor costs (Mosher and Crawford 1932; Slattery 1940).

With these quality-of-life and economic enhancements, Morris Cooke saw a "cultural renaissance" in the making, a revitalizing force for the nation, amid fear that decay of the heartland would led to a "historic slide" downward (Christie 1983; *Rural Electrification News* 1936a; Stauter 1973).[18] Electrification did indeed improve the quality of life in rural areas, and it also helped greatly increase agricultural productivity. But Stauter (1973) asks, "Was it 'cultural renaissance' or just another step in seemingly inevitable homogenization of American life? Farm life lost much of its drudgery, to be sure, but it also lost some of its distinctiveness and character" (274). Moreover, the reduction in the need for labor, with dramatic increases in productivity and increased integration with urban areas, enabled more people to leave the farm. Consequently, electricity, instead of retaining the rural population, contributed significantly to increased migration to urban areas (Stauter 1973).

In sum, what eventually transpired was at odds with what was envisioned. Consider the prediction of the governor of Pennsylvania, Gifford Pinchot, a champion of rural electrification: "Long distance electrical transmission is to be the basis of the new economic and social order" (Christie 1983, 73). None of that came to pass—reality did not accord with what Carey and Quirk (1970, 423) call expectations of "electronic sublime."

SYSTEM

Gains   Urban power plants also gained from electrification, despite their resistance. In the late 1930s, REA-assisted projects made about 67 percent of their total wholesale power purchases from private utilities. In the next decade, 1940–1950, their annual purchases from utilities increased twentyfold, from $2.5 million to $50 million (Wickard 1950). Also, manufacturers of electrical appliances profited from the expansion of their markets to rural areas. Poor's Industry and Investment Surveys reported:

The electric washing machine industry's unit sales volume will top that of 1935 by close to 30 percent. Sale of ironers will also establish a new peak. The demand for household electric refrigerators continued strong. For 9 months of 1936, sales showed a gain of approximately 30 percent over the corresponding 1935 period. Radio sales continue upward. (*Rural Electrification News* 1937, 3)

Rural electrification not only created an immediate market for electrical equipment, but it also created future markets by increasing farm income

(Dieken 1936). As Childs (1974) put it: "It was a golden market, a bonanza, that beckoned private enterprise" (81).

It is conceivable that REA could have developed its own generation capacity. It even threatened to do so. But that would have increased the complexity and cost of rural electrification. In actuality, the wholesale supply of the electricity by utilities to rural cooperatives was mutually beneficial (Zinder 1936).

Travails   In the light of the success of New Deal policies and programs, it is easy to overlook the political opposition they had aroused in the beginning. In the case of REA, the opposition was extraordinarily fierce because it helped create potential competitors to private enterprise—namely, utilities. The time and resources that the different stakeholders and their allies devoted to the political struggle resulted in an opportunity cost for the political system, if nothing else.

NEW CONCEPTIONS FROM THE MARGIN

We have much to learn from the arguments advanced by the proponents of rural electrification—both in terms of what to emulate and what to avoid. They had two sets of arguments: (1) reasons why utilities were wrong about the costs of rural electrification, and (2) reasons why rural electrification would benefit rural areas.

COSTS OF RURAL ELECTRIFICATION

The utilities' skepticism about the economics of rural electrification was anchored on two considerations:

1. Cost of constructing rural lines was very high.
2. The potential demand was too low.

In effect, from their standpoint, the math did not add up. The proponents thus countered the utilities' arguments as follows:

1. The utilities' cost estimates were based on the principle that each rural line should pay for itself, which is deeply flawed.
2. The utilities' estimates of potential demand were based on the assumption that rural households would use electricity for a "sixty-watt bulb"

(i.e., a handful of applications) for only a few hours a day, which reflected a limited understanding of the mix of domestic activities and working in rural life.

On the first argument, they went on to undertake recentering-on-reversal, pointing out that the notion that each rural line should pay for itself was a product of a linear conception of the rural network as an extension of the urban system. Herein, the former was reduced to an appendage of the latter (figure 4.1). The linearity of this conception, which in modernity has a ready hold on our minds, obscures other possible linking relationships between adjoining areas. The proponents decentered the urban areas and center-staged the earlier marginalized rural communities.[19] In this changed mindscape, lateral connections *among* rural communities gained salience.[20] Correspondingly, they developed the area coverage approach, wherein the urban-rural relationship is conceptualized very differently—the rural network as a complement of the urban system (figure 4.2). What is truly remarkable here is that the proponents not only undertook recentering-on-reversal on the conceptual plane, but they also demonstrated its power on the ground by building out successful rural electricity networks in accordance with the area coverage approach. Appropriately, upon proving the expert opinion wrong, Cooke

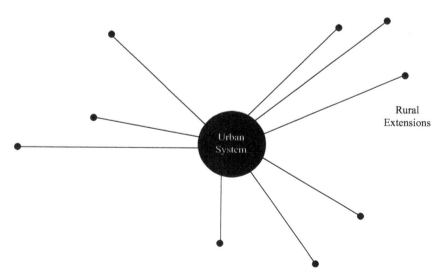

Rural
Extensions

Urban
System

4.1 Utilities' view of rural electrification: Appendage of urban system.

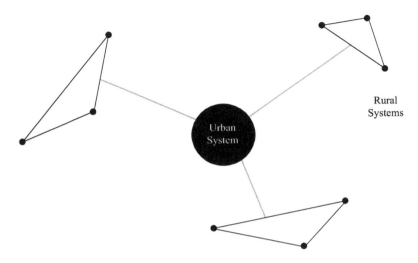

4.2 Proponents' view of rural electrification: Complement to urban system.

noted that REA-assisted cooperatives brought about "rural electrification in spite of the expert" (quoted in Christie 1983, 184).

On the second argument, rural dwellers from the outset started using electricity in myriad, often innovative, ways. The proponents, for their part, proactively worked to facilitate use and build up demand. For instance, REA provided loans for inside wiring and appliances, held demonstrations of electric machinery and appliances, and engaged in outreach activities. With the success of rural cooperatives, they showed that utilities, viewing rural demand through the urban lens, failed to realize that a rural household is not only a place of residence, but also a place of production.[21] In the words of John Carmody, who succeeded Cooke as the head of REA, the utilities "couldn't see the rural market."[22]

BENEFITS OF RURAL ELECTRIFICATION

The proponents sought to stem the depopulation of rural areas. To that end, they saw two broad benefits of rural electrification:

1. Enhancement of rural quality of life
2. Enhancement of rural economies

These enhancements, in their view, would reduce rural-urban disparity.

Electrification did enhance the quality of rural life—a point we need not belabor here. But that alone could not stanch outmigration to urban areas.

Proponents' expectations were thwarted by problems in the arena of economics, which they did not anticipate. Broadly, they were twofold:

1. Electricity did reduce drudgery and increase the efficiency of the farm operations, as they had anticipated. But then it also greatly reduced the need for labor, which they had not considered.
2. Electricity did enable urban enterprises to relocate operations in rural areas, as they had anticipated. But the economic decentralization that occurred paled in comparison to the other changes enabled by electrification, which they had not considered.

The proponents of rural electrification got blindsided because of their rural-centered view of electricity. Black (1962) explains how a metaphor can organize our view of a multidimensional object of interest:

Suppose I look at the night sky through a piece of heavily smoked glass on which certain lines have been left clear. Then I shall see only the stars that can be made to lie on the lines previously prepared upon the screen, and the stars I do see will be seen as organized by the screen's structure (41).

Translating Black into our domain of interest, the problems of rural areas are the clear lines, which organized the proponents' view of electricity. They saw in electricity the potential for solving rural problems. They were correct, as electricity could indeed help enhance the quality of rural life and enable decentralization.[23] But they were correct in only a small way. Their rural-centric view obscured the bigger reality: electricity is a protean technology that enables myriad things. Correspondingly, it is polysemous—different people take it to mean different things (Lieberman 2017).

The proponents' rural-centered view was severely constrained, centered as it was on what country dwellers would do with electricity. For instance, Slattery (1940) waxed enthusiastic about its decentralizing potential: "Rural area electrification and high-voltage transmission have put the equivalent of a large or small steam or gas engine at the command of the farmer" (96).[24] This view is centered on the "engine at the command of the farmer." It is a nodal view, focusing on nodes of the system. If we employ the analytical strategy of recentering-on-reversal and decenter the "engine at the command of the farmer," we can easily see what the

rural-centric view of rural electrification obscured—"high-voltage transmission." Now, if we center-stage "high-voltage transmission," we have
the relational view, which focuses on connections between the nodes,
allowing us to see what actually transpired.[25]

Electrification integrated rural areas into more tightly with the metropolitan economy. While electrification enabled the proponents to implement
their envisioned solutions for rural areas, it did not stem depopulation.
On the contrary, quite the opposite happened. Proponents had failed to
consider that in the larger arena of integrated metropolitan economy,
their own efforts to shore up rural areas would be dwarfed by other forces
at play. They were limited to a localized view, one too optimistically
biased toward local solutions, when they needed a global network view,
with its attendant implications—both positive and negative.

Looking back, the process would have been better served if the rural-
centered view of the proponents also had been subjected to recentering-
on-reversal, as doing so would have expanded the discussion in the
direction of what eventually transpired. The utilities would have been
prime candidates for undertaking such an exercise.

Since the utilities failed to rise to that level, our analysis suggests
that the proponents would have been better off if they had performed
recentering-on-reversal on their own rural-centered view. Doing that would
have expanded the range of their thinking to possibilities that were counter
to their hopes and intuitions. But that would be asking too much of under-
dogs engaged in a tough battle with much more powerful adversaries,
when the general sentiment, including those of sophisticated observers
of technology, celebrated electricity, including centralized power generation and long-distance transmission (Lieberman 2017). This, however,
is not inconceivable because for its proponents, electrification was only
a vehicle for their larger goal—the betterment of rural areas. To forestall
such slippage, perhaps recentering-on-reversal should be part of the standard procedure employed for policy formulation.

We happen to be ending with a rather tough critique of the proponents of rural electrification. As a matter of fact, they actually did pretty
well. They subjected utilities' urban-centric view to recentering-on-reversal,
and, moreover, they went on to construct self-sustaining electricity networks, improving the quality of rural life. The question with regard to

performance of recentering-on-reversal on proponents' own view arises because they lost on the bigger front—electrification facilitated further depopulation of rural areas as opposed to stemming it. But then, this is a coldly analytical exercise.

We often find ourselves taking a tough view of social compacts we cherish. The analytical strategy of recentering-on-reversal makes us think of things that we were committed to in new ways. In the case of proponents of rural electrification, recentering-on-reversal on their position might have served them well, giving them an opportunity to formulate interventions to forestall and mitigate movements in undesirable directions that they failed to anticipate.

5

TELEPHONY

In 1900, *Scientific American* reported a curious telephone line in use between Anderson, Pendleton, and Ingalls in Indiana. Intriguingly, it used the top wires of rows of barbwire fences, with gaps bridged with galvanized wires. Yet it provided reasonably good connections: "The line, it seems, is not an experiment, but it is in active daily operation with four regular subscribers, and it gives a service which our correspondent assures us compares well with the lines of regular companies" (*Scientific American* 1900, 196). C. Alley, the creator/owner of this fourteen-mile line, claimed that it was the only barbwire-based line in existence (*Scientific American* 1900). In reality, Mr. Alley's line was not the only one, but it was the first to come to the notice of *Scientific American*, whose prestige lent it cachet. For instance, two years earlier, in the *New York Times*, a two-paragraph article "Fences for telephone lines," reporting on plans of farmers of Washington Valley in northern New Jersey, said that "they had heard of a primitive line that is in use in the southern part of the State, and intend building one of their own if it be found practicable" (*New York Times* 1898, 4). Further, two years after the *Scientific American* report, another article in the *New York Times*, a reprint of a story in the *Butte Inter Mountain*, situated the creation of a barbwire-based line in Montana this way:

Fort Benton's latest effort is a barbed-wire wire communication. Being by instinct and association cow people, they resented the genesis of the barbed wire, and when it was discovered that it was one of the evils that came with the railroad and threw the cowpuncher, the bull-train, and the river streamer out of the game, they decided to take a material view of the situation, and the result is that they are preparing to have a telephone exchange which will take in every ranch from the Missouri River north to the Canadian line and south to the Highwood Mountains. (*New York Times* 1902d, 18)

Soon the practice spread, often inspired by the *Scientific American* report, in Colorado, Iowa, Kansas, Nebraska, New Mexico, Minnesota, Ohio, Pennsylvania, South Dakota, and other states (Kline 2000).[1]

Such lines were the butt of urban jokes. Ironically, even the independent telephone companies, formed to serve rural areas neglected by the Bell Telephone Company, made fun of barbwire-based lines, calling them "squirrel lines." Their industry publication *Telephony* reprinted local newspaper reports of such lines in its humor section (Kline 2000).

Whether or not barbwire-based lines were a joke is a matter of perspective, as we will see.

EXPANSION OF TELEPHONY

In May 1877, the Bell Telephone Company entered the market, offering to lease a set of two telephones for $40 per year for business use and $29 for nonbusiness use (Brock 1981).[2]

Sensing a competitive threat to its profitable intracity telegraph service, Western Union Telegraph Company entered the telephone business in December 1877, even though its legal standing vis-à-vis Bell's patents was tenuous. Western Union's entry set off an expansionary dynamic, with both companies rushing to enter new markets to secure first-mover advantage. Further, since the value of a network grows with the number of subscribers, both set about expanding their new networks as soon as possible (Gabel 1969).[3] But then, in November 1879, Western Union sold its telephone networks to Bell in a complex deal, exiting the telephone business under the pressure of patent infringement lawsuits and fear of a hostile takeover of its core business—the telegraph (Brock 1981; Hochfelder 2002).

After the exit of Western Union, Bell had a patent monopoly until 1894, when its key patents expired. Yet, Bell continued to set up service in all "desirable" locations[4] to forestall competition on the expiration of its key patents, as its 1880 annual report explains: "With a thorough occupation of the principal cities and towns . . . the danger of competition with our business from newcomers seems small" (Brock 1981, 103).

At the end of 1877, when Western Union entered the market as a competitor, the US had 9,000 telephones. At the end of 1879, when Western Union exited the telephone business, the number had reached 31,000 (Bureau of the Census 1976). Even as a monopoly, Bell continued its network expansion, tripling the number of telephones between 1880 and 1884, and then doubling them between 1884 and 1893 (Fischer 1987). At the end of 1894, when Bell's key patents expired, the US had 285,000 telephones (Bureau of the Census 1976).[5]

INEQUITIES AND INEQUITIES THAT MATTER

Bell's strategy had a blind spot—a limited conception of what constituted desirable locations. At the end of its patent monopoly, half of its telephones were in cities with over 50,000 people, aggregating to 18 percent of the US population. On the other hand, only 6 percent of its telephones were in communities with fewer than 10,000 people, aggregating to 71 percent of the US population (MacDougall 2004). In effect, the development of telephony thus far was largely an urban affair.

Bell ignored rural communities, seeing no economic benefit in serving them. In its estimation, not only was the cost of laying rural lines prohibitive, but rural people's telephony needs were limited. When it came to the social benefits of rural telephony, Bell executives belittled them as "the sentimental side of the question" (MacDougall 2004, 129). In general, for Bell, the very notion of rural telephone service was a "lost ball in tall grass" (Blalock 1940, 466).

While Bell was steadfast in its stance that urban-rural inequity did not matter, rural dwellers felt otherwise. "Farmers were, in fact, an 'ornery' lot. Many of them demanded this technology when its vendors said it was inappropriate for them" (Fischer 1987, 16). Bell routinely turned down petitions from small communities, with a response that a critic characterized as: "No, your town is too insignificant for us to consider" (MacDougall 2004, 127).

In this way, Bell's policies created a "reservoir of unsatisfied demand" (Brock 1981, 107). In the period prior to the expiration of Bell's controlling patents in 1894, farmers created jerry-rigged telephone networks that were illegal or of doubtful legality (Atwood 1984; Fischer 1987). In response, Bell expended its energies suing these "wildcats" as opposed to expanding its service; in the process, it infuriated people in unserved areas with what they saw as "dog-in-the-manger policies" (MacDougall 2004, 127).

Clearly, the inequity mattered to rural people, as they took great trouble to ask for service or create telephone connectivity themselves.

BOTTOM-UP INITIATIVES COMPEL ACTION

The proponents broke the stalemate not by way of argument, but action.

After June 1894, upon the expiration of Bell's patents, bottom-up entrepreneurial activity saw unbridled flourishing with the emergence of independents—telephone companies independent of Bell. Broadly, they were either profit-motivated entrepreneurial enterprises or cooperative endeavors such as subscriber-owned mutuals and farmer cooperatives. The entrepreneurs were after a "share of the boodle" that Bell had long monopolized (Brooks 1976, 104). For them, rural areas represented virgin markets, in contrast to urban areas where Bell was dominant. Cooperatives, on the other hand, were self-help endeavors.

By the end of the year, independents were serving 15,000 phones— about 5 percent of the total market; and they grew, reaching market shares of 19 percent in 1897 and 44 percent in 1902 (Brock 1981). With the rising threat of independents, Bell was forced to begin serving rural areas. The ensuing competition fueled network growth, with a ninefold increase in number of telephones per capita between 1893 and 1902, eclipsing the twofold increase in the previous nine years (Fischer 1987). In 1902, the total number of telephones in the US reached 2,371,000 (Bureau of the Census 1976).

Around the turn of the century, the independents also started making inroads in urban areas, and now towns often had companies competing for their business. In 1905, Iowa alone had 147 towns with two or more competing companies. Since the competitors' networks were not interconnected, businesses had to subscribe and advertise "both phones," given that residential customers typically subscribed to only one telephone service. Consequently, telephone networks often developed along class lines; for example, in Minneapolis, the older Bell network served the elite, while the newer Tri-State Telephone Company brought new subscribers of lesser means into the fold with its lower rates (Brooks 1976).[6]

By 1907, the independents had captured a 49 percent market share, when the total number of telephones in the US was 6,119,000 (Brock 1981; Bureau of the Census 1976). To stem this tide, Bell's division head, Thomas Doolittle, and president, Theodore Vail, strategically deployed long-distance telephony. Even though long-distance telephony was still technologically embryonic and had a doubtful future, both on technical and commercial grounds, the company started interlinking the Bell telephone networks, which were local in scope, into a national system. Against the doubters within his own company, Vail argued: "I take it that it is extremely important that we should control the whole toll-line system of intercommunication throughout the country. This system is destined, in my opinion, to be very much more important in the future . . . We need not fear the opposition in a single place, provided we control the means of communication with other places" (Brock 1981, 118). This system building strategy paid off big time in the long term.

However, in the short term, this move did not affect the independents, as the traffic was almost entirely local. Furthermore, the telegraph was ubiquitous and economical. Contrary to Bell's expectations, regional calling within a 100-mile radius turned out to be the arena of decisive competition. A local telephone company could increase the value of its service by interconnecting with telephone companies in nearby areas, expanding the calling universe of its subscribers. The independents started making such interconnections among themselves, while Bell focused on building out its national long-distance network. Later, Bell, realizing its mistake, entered this arena in a concerted way. The company's systematic approach secured a decided advantage over the loosely coordinated and piecemeal approach of the independents. Moreover, it started enticing susceptible independents into interconnection arrangements as sublicensees. Other independents called out sublicensees as sellouts to the "octopus"— their epithet for Bell, which also had many tentacles, metaphorically speaking, with its extensive economic and political reach. But for the sublicensees, such arrangements were a way out of a tough competitive situation and a lifeline for economic survival, especially when Bell offered to vacate the area where it was in competition with them (Mueller 1997). Gradually, Bell secured a dominant position region by region, and later its national long-distance network made its overall system unassailable.[7]

Around the same time, Vail started talking about universal service in Bell's annual reports, starting with the 1907 one. In this way, Bell's celebrated slogan—"One system, one policy, universal service"—appeared in the 1909 annual report (Mueller 1993). Vail meant something very different by "universal service" than what we understand it to be today. He was clearly not calling for service for everyone, as he was against rural telephone service. Instead, he was advocating for a unified system, operated by one entity, that allowed any subscriber to talk to any other one, which was not possible at the time because of lack of interconnection among competing networks. Basically, Vail was making a case for a Bell takeover of the industry to "unify the service," seeking the government's permission to acquire independents in spite of antitrust laws (Mueller 1997).

At first, the government's response vacillated between impedance and acceptance. Gradually, however, it became decidedly opposed to Bell's acquisition of the independents. Its threat of antitrust action led to the Kingsbury Commitment of 1913, in which Bell agreed to stop acquiring independents and also to interconnect with them. This saved the independents, but it also reduced them to appendages of a Bell-centric, integrated system (Barnett and Carroll 1993; Brock 1981).[8] At that point, the US had a total of 9,543,000 telephones (Bureau of the Census 1976). Subsequently, in the interwar years, the network grew gradually. In 1945, the US had a total of 27,867,000 telephones and 46.2 percent of households had service (Bureau of the Census 1976).[9]

SYSTEM-BASED EXPANSION OF UNIVERSAL SERVICE

By 1965, under the Bell-centric, integrated system framework, the number of telephones in the US had reached a total of 93,656,000; and 84.6 percent of households had service (Bureau of the Census 1976). Thereafter, the regulators started making deliberate use of intrasystem cross-subsidies—the transfer of monies from profitable long-distance service to subsidized local residential services—to achieve universalization of the network, and with that, the dynamic of network expansion changed.

The discussion on cross-subsidies started in the 1920s, but only in 1947 did the Federal Communications Commission (FCC) put in place a mechanism for enabling it—namely, the *Separations Manual*.[10] The manual laid out the method for allocating local loop costs between local service, which uses only the local loop, and long-distance service, which uses both the long-distance network and the local loop. Interestingly, when the recommendations were instituted, the regulators did not see that as a means of advancing universal service. As late as 1965, less than 3 percent of the local loop costs were recovered from long-distance service.

Starting in 1965, however, the regulators deliberately began moving money from long-distance service to subsidies for local service, and the amounts increased over the years. This process proceeded in accordance with political understandings and technocratic considerations of the powers that be. The politicians wanted affordable local telephone service and Bell, understandably invested in the preservation of the Bell-centered system, embraced universal service as a justification for preserving the status quo. Gains from the declining costs of long-distance technologies provided a source of money to subsidize local service, and the FCC essentially provided the technocratic expertise needed for its collection and disbursement. In this way, by around 1980, telephone penetration eventually reached 90 percent of US households (Belinfante 2006; Jayakar and Sawhney 2004; Mueller 1997; National Governors Association 1988).

In sum, in the case of telephony, the argument for the extension of service to rural areas was primarily made in action, wherein those desiring it simply went ahead and built networks for themselves—illegally at first, but then lawfully after the expiration of Bell's patents. While Bell could see the social and economic need for telephone service in rural areas, it could not fathom its intensity. Clearly, Bell misread rural demand on both counts—social and economic. The mushrooming of bottom-up activity was a clear indication that inequity mattered to those who were affected by it. Moreover, it also established that the demand for telephone service in rural areas was sufficient to make the investment economically viable. More broadly, the resulting network provided an adequate foundation for the subsequent development of universal service, which was a gradual process.

GAINS AND TRAVAILS

At the turn of the twentieth century, *Drover's Journal*, published in Nebraska, noted that "no modern invention has so thoroughly revolutionized rural communities, as the telephone" (Dilts 1941, 32). Such proclamations of revolutionary change are typically celebratory, as was the case here. Even when there is much to celebrate, as we will see, our analytical project prompts us to look beyond the gains to the travails—a look that is deep and long term (table 5.1).

INDIVIDUAL

Gains What Bell had dismissed as "the sentimental side of the question" was a major existential issue for rural dwellers, who had to traverse vast distances on a daily basis. Testimonials on the need for connectivity abound, including the following: "When bitter winter days housed us in we had only one contact with the world. That was the telephone" (Nordyke 1938, 24); "It was heavenly to have the telephone. Talking to folks is next thing to seeing them" (Pound 1926, 32); and "an ever ready friend" that could "make your farm house the center of a city" (Atwood 1984, 82). These comments, as sentimental as they might sound, speak to rural dwellers' gains from access to telephone service.

Table 5.1 Universal service: Gains and travails

	Gains	Travails
Individual	Reduced barriers of distance for social communications Reduced barriers of distance for business communications Enables calls for emergency assistance	Dividends of connectivity are dependent on power differentials between the parties connected Consequences of the scaling-up of telephony from area networks to a global one were not understood; scaled-up networks engender asymmetrical extralocal relationships
System	Revelation of untapped latent demand in rural areas it had failed to fathom Quickened pace of technological advances	Caught off guard by the emergence of competition in rural areas Access issues attracted federal scrutiny

On the economic front, too, rural dwellers have a greater need for telephones, given the rural distances, than their urban counterparts (Mulrooney 1937; Pound 1926). They also have a higher intensity of use, as a rural homestead is both a place of residence and a center of production (Fischer 1987). For instance, a study on the incoming and outgoing calls of twenty-seven Iowa families had the following breakdown: business (44.6 percent), social (35.2 percent), household (11.1 percent), and miscellaneous (9.1 percent) (Holland 1938).[11]

Travails Neighboring farmers' desire to somehow connect with each other motivated them to put together jerry-rigged networks. Once rural telephony became a business, the independents shifted their focus to connectivity among small towns. By building these short-haul toll lines, the independents increased the value of the services they offered (Mueller 1993). As Phillips (1939) noted, "even the value of the town telephone is largely made of its ability to contact the farming trade area surrounding our small communities" (24). These short-haul lines, in other words, facilitated interactions between parties that already had long-standing ties (e.g., a farmer and the supplier of agricultural implements in the nearest small town). Leveraging their position as area hubs, the independents went on to establish connectivity with city businesses interested in communicating with their customers in the countryside. This was a case of "the periphery advancing on the center" (Mueller 1993, 360).

This worked fine until the center, Bell, was provoked into action. Then the full consequences of connectivity, which is a double-edged sword, started to kick in.[12] Bell first started building its long-distance network, and later realizing that the independents were growing in the 100-mile-radius regional calling market, it moved into this arena. Employing a systematic approach, including enticing many independents into sublicensee arrangements, Bell prevailed over the independents. Federal intervention and the subsequent Kingsbury Commitment obligating interconnection—albeit at terms unfavorable to the independents (see Mueller 1997)—and the cessation of acquisitions saved the remaining independents but reduced them to appendages of a Bell-centric national system.

In sum, the farmers, and then the independents, deployed telephonic connectivity in ways that empowered them. At first, they could do so relatively simply because they were operating in the isolation of rural areas, which were not of much interest to Bell. Later, when Bell did become interested, the arena greatly expanded. While the independents continued to deploy telephonic connectivity for their own empowerment, in the expanded arena, new actors who pursued their own goals were also in play. Moreover, these bigger and more powerful actors, especially metropole-based corporations, did not deploy telephonic connectivity to facilitate and strengthen existing relationships. They instead deployed connectivity to forge new relationships that were mostly metropole-centric and far more asymmetrical than the ones of small-town businesses and nearby farmers. The collateral damage of this metropole-centric refashioning was the fraying, if not sundering, of existing relationships, which were local and far less asymmetrical. Now connectivity was manifestly a double-edged sword.

As a general principle, the dividends of connectivity depend on the power differentials among the parties connected (Mulgan 1991; Samarajiva and Shields 1990a, 1990b). In the period when the independents ruled the roost in the rural areas, a period when Bell ignored the countryside, the power differentials between the area hubs and the surrounding communities were much smaller, relatively speaking. The dividends of connectivity, which were significant in the local context, stayed local. With the formation of a nationwide system, rural communities were now embedded in a system marked by far greater asymmetry, and consequently, the dividends of connectivity overwhelmingly accrued to distant parties.

SYSTEM

Gains   Bell also profited from bottom-up rural telephony initiatives, as they revealed untapped latent demand that it had failed to recognize. Furthermore, the resulting competition fueled technological advancement. For instance, Bell's reluctance to move from manual switching systems, which it had painstakingly perfected, to mechanical switching systems, which the independents had pioneered, lasted only until competitive pressure compelled it to do so (Mueller 1989). On another front, Bell pushed the development of long-distance technologies to secure a decisive competitive advantage (Garnet 1985; Lipartito 1989). These technological advancements very likely would have occurred even if Bell did not face competitive pressure, but not at the pace they did under that pressure. Moreover,

over the long haul, Bell not only emerged as the victor, but it also enjoyed a nationwide, integrated network. In this new order, the independents that remained were reduced to appendages of the Bell system.

Travails   Bell was caught off guard by the emergence of competition in rural areas. Furthermore, this competition grew vigorously, with independents' market share reaching 49 percent in 1907. Bell was at a loss about how to contain this until its long-distance strategies started to work.

Furthermore, competition in rural areas brought the Bell system under close federal scrutiny. It had avoided antitrust action by making the Kingsbury Commitment of 1913, but eventually, another antitrust action led to the breakup of the Bell system in 1982.

CONCLUSION

Parallels in the development of rural electrification and rural telephony are striking. The utilities conceived of rural networks in similar ways, and the proponents of rural service critiqued their conceptions in similar ways. Furthermore, proponents' actions on the ground, not their arguments, settled the debate. However, they differed in one significant way: public policy intervention. While the rural electricity cooperatives could erect their own distribution networks, they needed electricity from utilities' power stations since they lacked the scale to generate power on their own cost-effectively. Here, they had to make a case for government intervention, requiring them to clearly articulate their critique of private utilities' approach. Rural telephony initiatives, on the other hand, were not encumbered by such considerations, at least initially, and had a much freer hand. Consequently, they clearly articulated their critique of Bell's approach much later in the process, when a public policy intervention was needed. Initially, their moves were mostly in the realm of action.

Bell was dead against rural telephony, firmly believing that it was economically unviable. On the cost side, it saw the construction of lines across the rural expanse to be prohibitively expensive. On the demand side, it saw limited scope for building up demand, as the population density was low and a typical user's needs seemed to be limited. When estimating construction costs, Bell took rural lines to be extensions of the urban network. In such a conception, the rural geography loomed large—here, the first step necessarily had to be lines from the urban network to distant settlements. In effect, geography as the determinative factor was intrinsically built into this perspective. From this standpoint, it was difficult to see things in another way, reducing the "ornery" farmers' demands for service to a nuisance born of a pipe dream (Fischer 1987, 16).

Situated on the margins as opposed to the center, farmers saw things very differently. With their first priority being local connections—to neighbors and nearby businesses—and not the long link to the metropole, they instinctually engaged in a recentering-on-reversal move.[13] When one decenters long links and center-stages local connections, geography is no longer the determinative factor, opening up a new way of thinking—one that prompts action, in marked contrast to the paralysis induced by the urban-centric view.[14] Moreover, network builders have much greater freedom of action, since now the city is no longer the inevitable starting point for network development and any useful link is a valuable one. What the independents needed was the *means* for making such connections. Given the elemental state of the technology—wires and manual switchboards—this was not a major hurdle.

On the longer arc of rural telephony's development, the early bottom-up initiatives made a critical contribution by revealing and satisfying untapped latent demand. In its limited view, Bell saw the unauthorized activities of entrepreneurial spirits that strung up jerry-rigged networks as threats to the established order, treating them as a nuisance that needed to be squashed and, at the same time, also downplaying their significance. Looked at in another way, they represented unsatisfied demand, which the independents took upon themselves to satisfy. Once the independents demonstrated the existence of this latent demand, Bell found a new willingness to extend its service to rural areas. Upon entering rural areas, it expanded its presence rapidly, which in turn spurred independents to expand their networks.

Initially, when rural networks first sprouted, regional connectivity was not a major factor. Much like the Anderson-Pendleton-Ingalls network that *Scientific American* first reported on in 1900, they tended to be very local affairs. Later, when independent networks grew, their subscribers' needs called for connectivity to business associates, acquaintances, friends, and family in the cities. Since Bell refused interconnection with its urban networks, the independents started entering urban markets, resulting in cities with competing telephone companies in the same area, where subscribers of one telephone company could not talk to those of another. On its part, Bell sought to stem the rising tide of competition by developing long-distance lines that integrated its local networks into a national system. However, at that time, users' needs were mostly regional— nearby towns and the nearest big city. While Bell focused on building out its national system, independents started interconnecting with each other to satisfy users' need for regional connectivity. Seeing their success, Bell realized its mistake and started establishing regional connectivity by enticing many independents to interconnect with its system as sublicensees. These sublicensees essentially became appendages of the Bell system. Subsequently, with the Kingsbury Commitment and other regulatory actions, other independents also became appendages.

The independents, in their celebration of telephonic connectivity, did not see that connectivity has different consequences at different scales. When they created their networks, they were local and isolated. Connectivity at this scale was productive in ways that worked for them, as it facilitated communications within existing personal, social, and business relationships. With the expansion of their networks and interconnection with other networks, the nature of connectivity and its consequences changed qualitatively. In this new arena, connectivity was still productive, but in ways where the extralocal overwhelmed the local. As a result, the independents lost their autonomy. What about the larger rural community?

Unlike electrification, no grand rhetoric warranted the proponents' efforts to develop rural telephony. They often touted the benefits of rural telephony, but, in marked contrast to electrification, proponents rarely presented or even thought of telephony as a vehicle for regenerating rural areas. Their efforts were scrappy and pragmatic, servicing immediate needs in different ways in different places—there was no sustained rural development framework per se.[15] Still, as in the case of electrification, proponents focused on the benefits, giving little or no thought to the potential downside. In reality, telephonic connectivity became another strand in the tighter integration of rural areas into the metropolitan economy, with consequent loss of their autonomy, much like that of independents themselves.

What is curious from the standpoint of our analysis is that the independents scoffed at the barbwire-based lines of the sort strung by Mr. Alley in Indiana. In their arrogance, they pretty much did to people like Mr. Alley what Bell had done to them—something that they had resented. From the standpoint of our analytical project, the independents' reaction to a barbwire-based network calls for reflection and analysis.

We moderns have an easy proclivity to downplay the oddities at the margins of established order, even recently established ones, as in the case of the independents. This could be a defense mechanism—a wishing-away of the anxieties of destabilization, and refuge seeking in the stability of the familiar order (Gehlen 1980). But then, such downplaying could also be a missed opportunity for leveraging recentering-on-reversal. The reality is that recentering can continue ad infinitum, as every order provides only a partial understanding. Even a new order born of a recent recentering, such as the one the independents performed, could in turn be subjected to further recentering-on-reversal. Accordingly, although all oddities at the margins of the established order cannot be pivots for recentering, we should keep in mind that any of them could have that potential—and therefore refrain from downplaying them. Eventually, at some point in time, someone will perform a recentering-on-reversal.

Against this backdrop of a larger discussion on recentering-on-reversal, let us do what the independents did not do, and see what we can learn. If we center-stage the barbwire-based lines, what do we see? We see the existence of unmet demand and a novel way of satisfying it. We then have the question of whether there is something qualitatively different here. With regard to unmet demand, in our estimation, it seems to be a matter of degree, as opposed to something qualitatively different. Independents understood the unexpected existence of unmet demand, as it was their own raison d'être—the barbwire-based lines basically suggested that demand was wider than what they had anticipated. Further, we find no reports of independents refusing to extend service to more peripheral parts of rural areas where the barbwire-based lines popped up, in marked contrast to Bell's refusal to serve rural areas.

So, it is likely that in due course, the independents would have expanded their networks to areas served by barbwire-based lines. With regard to the mode of networking, however, barbed-wire connection was truly novel, capitalizing as it did on preexisting assets—barbwire fences—on small patches of land. At the same time, it would be foolhardy to replicate this in areas that did not have preexisting barbwire fences, given that regular telephone wires are cheaper and carry transmissions much more clearly. Furthermore, this mode of networking cannot be easily scaled. All in all, in our estimation, the barbwire-based lines were then basically temporary solutions, akin to military field telephones.

In effect, the independents' downplaying of the barbwire-based lines did not leave them vulnerable to competitive pressure or undermine the development of telephony in general. But the attitude was problematic, both at the human level and the analytical level, where the story need not stop after a single recentering-on-reversal—no matter how spectacular.

Of all the systems covered in this book, telephony is a particularly edifying case in multiple ways. It shows a spectacularly successful instance of recentering-on-reversal, the one performed by the independents. It also shows psychological barriers to reversal, most pointedly in the independents' dismissive attitude toward barbwire-based lines. In addition, it reveals that all oddities at the margins of the established order cannot be pivots for recentering-on-reversal (e.g., barbwire-lines). Yet they need to be examined, closely, and not dismissed. In sum, while the analytical strategy of recentering-on-reversal is not a silver bullet, it is a powerful tool that can deliver big in many contexts, if carried out with care.

6

PUBLIC LIBRARIES

On December 5, 1877, dedicating a branch of the Boston Public Library, Mayor Frederick Prince claimed, rather triumphantly, that the public library "has become so fixed in the affections of the people that it may now safely defy all opposition" (Boston Public Library 1878, 4). The mayor's assessment was generally accurate; by that time, the library had become part and parcel of urban life in New England and beyond. If there was disagreement—and there was, and still is—it had largely to do with the function of public libraries, not with whether they should exist at all.

With regard to the function of the public library, the specific points of contention have changed over the years. In the late nineteenth century, debate raged on how the public library should address the growing popular interest in works of fiction, especially among women. Should the library guide readers away from works of fiction or make them readily available? Today, the points of contention often relate to internet access in public libraries (e.g., the installation of filtering software for protection of children). While the specific points of contention change, the friction point is deeply rooted: the normative view of the stewards of the library versus the alternative perspectives within the public.

In this chapter, we examine the underlying dynamics of the long-running debate on the role of the public library. We then look for blind spots and see whether the analytical strategy of recentering-on-reversal could be helpful in identifying them.

EXPANSION OF PUBLIC LIBRARIES

The early public libraries in the US varied in accordance with their local specificities: For instance, Salisbury, Connecticut, founded a public library in 1810, the first in the nation, when a vote at a town meeting approved the use of public money to build on a gift of 150 books from a publisher;[1] Peterborough, New Hampshire, established its public library in 1833 with funding from the State Literary Fund, tax money that had proved inadequate for realizing the original intent—building a state university;[2] and Wayland, Massachusetts, received a $500 gift in 1847, prompting 208 of its residents to donate a total of $553.90 for establishing a public library.[3]

Many influential figures embracing "universal cultural brotherhood" saw the public library as the "destroyer of class distinctions, sectional antagonisms, and international ill will" (Ditzion 1947, 6). Francis Wayland, a prominent Baptist educator, spoke of the public library as an instrument for cultivating a virtuous culture, which kept decadence at bay.[4] Edward Everett, a Unitarian clergyman, thought that the public library would fill Boston's "noble system of public instructions" by aiding life-long learning (1851, 255). George Ticknor, a prominent leader of the library movement, holding illiteracy to be the germinator of social upheavals, championed the public library as a vehicle for promoting self-culture. In general, the elites saw the library as a means for countering "tendencies to dissipation" and producing educated and well-behaved workers (Ditzion 1947, 24). In this fashion, a medley of voices, distinct but converging, laid the groundwork for the development of public libraries in Boston and beyond.

In 1876, the US had approximately 300 public libraries with at least 300 volumes,[5] two-thirds of which were in the Northeast. In the following decades, the public library movement gained momentum. On the one hand, a number of states, starting with Massachusetts in 1890, passed legislation for state-level support of public libraries. On the other hand, the industrialist Andrew Carnegie began providing funds for the construction of libraries—the first Carnegie library opened in 1889. With these initiatives, public libraries spread rapidly until World War 1 (Prentice 2011). In

1913, on the eve of World War 1, the US had about 2,100 public libraries with at least 5,000 volumes (Kevane and Sundstrom 2014). Carnegie had funded the construction of about 1,700 public libraries, including some with fewer than 5,000 volumes (Mitchell 2018).

Interestingly, the Progressive Era's public library legislation and Carnegie's philanthropy were similarly motivated: the elevation of the working class, assimilation of immigrants, and advancement of civilization. In the interwar years, Carnegie did not fund library construction, focusing instead on librarian training (Anderson 1963). Support by the local and state governments for the expansion of public libraries continued, with the number of libraries having at least 5,000 volumes reaching about 5,600 in 1929, when the Great Depression started (Kevane and Sundstrom 2014). After World War II, federal funds enabled further expansion of public libraries (Prentice 2011). Today, there are about 9,000 public libraries (not counting branch libraries) in the US.[6]

While scholars largely agree about such facts, their interpretations diverge. Ditzion's early account, which gave much weight to social reformers such as Wayland, Everett, and Ticknor, is considered to be a "progressivist interpretation" in need of serious revision. The historian Dee Garrison (1979) provides one such revisionist interpretation, which, as we shall see, views librarians as closely aligned with the upper and middle classes.

INEQUITIES AND INEQUITIES THAT MATTER

Although the role of New England and New York elites is often exaggerated and rightly criticized by some historians (e.g., Garrison 1979; McCook 2001), there is no question that their home states were the harbingers of public libraries. Moreover, the dynamic set in motion by the New England and New York elites shaped the development of the public libraries all over the US—public libraries spread to the rest of the country in a westward movement.

This dynamic differed from what we see in the case of other systems—the inequities mattered far more to the providers of access, who were certain about their mission, than the intended beneficiaries, who were conflicted. For instance, in 1892, a labor organization in Pittsburgh petitioned the city council to return Andrew Carnegie's donation, considering the money to be tainted and the donation motivated by desire for self-glorification—a sentiment echoed in Detroit in 1901 and in Indianapolis in 1903. But then the communities did accept Carnegie's money. For his part, Carnegie mainly asked that communities provide land and a commitment to take care of the libraries in the future, an assurance that they readily provided.[7] The communities typically located public libraries, those funded by Carnegie and others, in the middle of the town, being points of local pride (Lerner 2009).

Here, we will focus on the conflicted nature of this process. We will start by understanding the New England and New York elites and their objectives. We will then look at how the intended recipients of their benefaction, mainly members of the working class, responded. Thereafter, in the section on gains and travails, we will analyze the upshot of this interaction between the elite-driven development of public libraries and the complex responses of the intended beneficiaries.

AMBITIONS OF THE BENEFACTORS

The founders and leaders of public libraries came almost exclusively from the ranks of what Garrison calls the "new gentry"—mostly urban, mostly white, and mostly male professionals, well-to-do individuals with literary affinities, businessmen committed to genteel standards, and others (Garrison 1979; Joeckel 1939; Prentice 1973; Shelton 1976). They shared a consonant worldview, which Garrison (1979) describes as follows: "The belief that America was a radical democratic experiment in government; the sense of urban crisis and chaos; the fear of immigrant intruders; the emphasis upon the family as guarantor of tradition; the discontent of women and labor; the hope that education would right the wrongs of poverty and crime; the hunger for education among the poor" (62).

In general, the elites were anxious about the huge changes underway in American society—changes that largely stemmed from mass immigration and industrialization. In this respect, the development of public library parallels that of universal education. In 1876, W. E. Foster, the director of the public library in Providence, Rhode Island, succinctly described the relationship between the public library and the public school as "two halves of one complete purpose" (Nardini 2001, 116).

In the case of universal education, as we have seen, its proponents championed the tax-supported public school as the "principal digestive organ of the body politic" for Americanizing immigrants, especially Catholics who were prone to setting up parochial schools (Strong 1963, 89). Likewise, the proponents of the public library championed it as an instrument for assimilating immigrants (Burgess 2013; Lerner 2009). Ticknor argued that public libraries would bring "them in willing subjection to our own institutions" (Harris 1975, 6). In the same vein, others argued that public libraries would dispel the "foreigner" of rebellion and disloyalty, which were seen as rooted in ignorance (Wadlin 1911; Wellard 1937, 56).[8]

Also, as in the case of universal education, which was taken to be a means of "reconciling freedom and order" (i.e., tutoring the newly enfranchised lower classes to respect the established system, especially property rights), public libraries were seen as a stabilizing force for the society at large, not just for immigrants (Kaestle 1983; Lerner 2009). Carnegie articulated this view very vividly when he said, "the result of knowledge [gleaned from libraries] is to make men not violent revolutionists, but cautious evolutionists; not destroyers, but careful improvers" (Harris 1975, 15).

The protests and strikes of the 1890s raised these concerns to a new level of urgency. About 14,000 protests erupted between 1880 and 1894, sparked by the railroad strike of 1877, which was crushed by local and state militias and federal troops. In the eyes of the New England and New York elite, protests, strikes, and lockouts by the working class, which included a large number of immigrants, were signs of impending social decay. According to an editorial in the *Library Journal*, this social decay could be prevented by increasing support for public libraries, which "furnish the most effective weapons against the demagogic ignorance. . . . Every book that the public library circulates helps to make . . . railroad rioters impossible" (*Library Journal* 1877, 395).

Conversely, the elites saw the public libraries as a means for creating productive and well-behaved citizens—in accordance with their conceptions (Lerner 2009). In this order of things, the librarian was the "teacher" and "missionary" to effectuate these ambitions (Nardini 2001). The stewards of public libraries, in short, were elitists with a beneficent face, motivated by varying degrees of altruism (understood in a particular way) and self-interest. And this affected how the rest of the society perceived public libraries from the outset.

RESPONSE OF THE INTENDED BENEFICIARIES

Notwithstanding the hopes and ambitions of librarians and their genteel supporters, the development of the public library was never a "vital working-class issue" (Garrison 1979, 49). People thought of the library as a valuable addition to their community, but not essential to daily life in the way that schools, hospitals, utilities, and the police were. In fact, the subtle paternalism of libraries repelled many people of the working class, whose long working hours did not leave them much time for reading books anyway.

The metaphors used by librarians themselves are quite telling. William Fletcher, president (1891–1892) of the American Library Association, characterizing the librarian as a cook following an old recipe for cooking a hare, noted that the initial step was elemental: "first catch your hare" (Fletcher 1904, 17). Works of fiction, in this thinking, were considered an effective means, the carrot, for "catching the hare" (Fletcher 1904, 17). In the same vein, Justin Winsor, president (1876–1885 and 1897) of the American Library Association, advised that once folks were in the library, "you must foster the instinct for reading, and then apply the agencies for directing it. . . . Let the attention be guided, as unwittingly as possible, from the poor to the indifferent, from this to the good, and so on to the best" (Winsor 1876, 64).

Furthermore, people of the working class were disaffected by the middle-class formality and "feminized propriety" of the library (Garrison 1979, xiii). In certain cases, people were even inspected for "a clean shave, a clean collar, and recent shine" before they were allowed access to the "better class" of literature (Kalisch 1969, 84). These measures turned libraries into uninviting and alien spaces for the working class. Harris and Spiegler (1974) characterize them as "inhospitable and cold for the man on the street" (264).

Interestingly, in this "feminized" space, the way that the late-nineteenth-century libraries dealt with women's growing interest in recreational reading—especially fiction featuring a new type of heroine, sensual, active, and defiant—is telling. At first, librarians met this demand for recreational reading with resistance, imposing strict censorship measures, closely monitoring reading habits, publishing guidebooks for good reading, and even circulating lists of so-called questionable authors. Only at the turn of the century, after tasting the failure of such measures, they finally gave up on clamping down "pleasant books" (Hill 1902, 13). They became overtly more accommodating while working to subtly place limits, such as implementing a "'two-book' system," whereby patrons could borrow two books concurrently if one of them was not a novel (Wiegand 2015, 80).

The critics were not against the notion of a public library per se, but rather the particular expression of it given by the elites. For instance, Eugene Debs, the five-time Socialist Party of America nominee for US president, called for public libraries to be in "glorious abundance when capitalism is abolished and workingmen are no longer robbed by the philanthropic pirates of the Carnegie class" (Garrison 1979, 49). It is also telling that when unions undertook worker education, they did not rely on the public library—they developed their own book collections (Garceau 1949).

GAINS AND TRAVAILS

Today, Americans are able to access the services of public libraries at more than 16,500 sites (including main buildings and branches). Moreover, the national average population-weighted distance for the closest public library is only 2.1 miles[9] (Donnelly 2015). In effect, public libraries, albeit with uneven resources, are widely accessible at the physical level. About half of the US population uses the public library in some way. Given our analytical project, we should not lose sight of the fact that half of the population does not use the public library at all, even when surveys show overwhelmingly positive assessments of the public library, including by many nonusers. With this in mind, and alert to the need to capture a fuller picture, we now take stock of the gains and travails (table 6.1).

INDIVIDUAL

Gains In 2017, about 96 percent of Americans lived within a public library legal service area, a little more than half were registered users, and

Table 6.1 Public library: Gains and travails

	Gains	Travails
Individual	Free access to books and other information sources Librarians often going the extra mile to provide information on government agencies and community organizations that could be of help Libraries as communal spaces Libraries helping the homeless	Part of the array of disciplinary institutions that seek to mold the individual Alienating environment in regard to class sensibilities Having to pay taxes whether or not one considers the library to be a public good
System	Libraries complementing the school system Libraries providing opportunities for the development of information technology skills, which are important for today's economy Libraries helping to dispel discontent by reducing ignorance and containing it by providing harmless entertainment	Only about half the population visiting a public library at least once a year A sizeable section of the population (about one-fifth) never visiting a public library

they made on average 4.2 visits a year (Institute of Museum and Library Services 2019b, 2020). Further, a Pew Research Center (2016) survey spotlights the high regard that Americans have for public libraries. Here, it is interesting to note that the majority of people who have never visited a public library thought that closing their local library would have a major impact on their community. Cleary, public libraries do a lot of good.

Individuals have free access to books and other information resources. In addition, librarians often go the extra mile to provide information on government agencies (e.g., child welfare) and community groups (e.g., community bicycle repair cooperative) that could address individuals' needs. Libraries also serve as valuable communal spaces—providing meeting rooms, among other things. Moreover, they have been at the forefront of issues related to homelessness, as the homeless gravitate toward them for shelter, restrooms, internet access, newspapers, books, and meeting friends.[10] Libraries help the homeless look for jobs and housing and interface with government agencies to provide food stamps and other assistance. Some libraries even hire psychiatric social workers and health and safety advocates, and also offer haircuts, meals, parking lots for shopping carts, and bus fare[11] (Fox 2015; Gunderman and Stevens 2015; Ruhlmann 2014; Simpson 2014).

Travails   Reports on the public library tend to overlook the nonusers. For instance, the reports of the Institute of Museum and Library Services, which administers the annual Public Library Surveys as per the Library Services Act of 2010, are full of statistics on collections, programs, services, and usage (e.g., Institute of Museum and Library Services 2018, 2019a, 2019b, 2020). Usage data, which is of interest to us, is provided on a per-user basis (e.g., book circulation per person) and broken down on the basis of location (city, suburb, town, rural) and the size of the population served. On a per-user basis, these numbers look good—but what about nonusers? Based on a 2016 telephone survey of 1,601 Americans (16 years and older), Pew provides demographic information on users (age, sex, race, education, and other characteristics). More important, though, it also provides "a portrait of those who have never been to libraries" (Pew Research Center 2016, 15). According to Pew survey data, 19 percent of Americans have never visited a public library. Who are they? They are likely to be male (24 percent), sixty-five years and older (26 percent), Hispanic (32 percent) or black (28 percent), educated up to high school graduation or less (29 percent), and living in households with income below $30,000.[12] Why they do not use the public library? This is our blind spot—one with consequences, as we will see in the next section.

We have two insights from the literature into what might be keeping people away from the library. One, the public library is part of the array of disciplinary institutions that mold individuals along lines that align with the system's imperatives.[13] It was established as "conservator of order," molding individuals in particular ways (Harris 1975, 14). Two, the sensibilities and ambience of the public library create an alienating environment for the working class (Garrison 1979; Harris and Spiegler 1974; Kalisch 1969).

It is important to note for our analysis that whether or not individuals use the public library, they have to pay taxes that support it.

SYSTEM

Gains   The public library complements the public school system, extending education to after-school hours for students still in school, and providing lifelong learning opportunities for former students who have entered the workforce. In our time, they provide people valuable opportunities to develop digital skills, which are important in our economy, helping them function as both workers and consumers.

The public library also helps dispel and contain discontent (Lerner 2009). It helps remove ignorance, which elites believed to the root cause of discontent (Ditzion 1947). If nothing else, it provides "a harmless form of entertainment" (Harris 1975, 3). For instance, during the temperance movement, its supporters saw public libraries as a means of keeping drinking men out of the alehouse—"an asylum for the inebriate" (Ditzion 1947, 23–24). In the same vein, reformers saw children's libraries as "wholesome alternatives to the street-corner and saloon" (Lerner 2009, 142). Today, the library provides access to various forms of electronic recreation.

Travails   As noted earlier, only about half of the population visits a public library at least once in any given year. Moreover, there is a sizable section of the population, about one-fifth, that has never visited a public library. On the one hand, lack of participation by a very large part of the population is contrary to the "library faith" (Garceau 1949, 50) in the centrality of public libraries in the democratic process. On the other hand, the fact that about one-fifth of the population remains totally beyond the reach of the library points to its shortfall as a disciplinary institution—perhaps from the standpoint of the system, these are the very people who need to be brought more deeply into the disciplinary regime.

SUCCESS AND ITS COMPLEMENT

Proponents of the public library never fail to underscore its special place in a democracy. Josephus N. Larned, president (1893–1894) of the American Library Association, went so far as to characterize the public library as the only institution capable of delivering a democratic future that is "safer and surer than any others that society can build hopes on" (Larned 1894, 4). In their preoccupation with the public library's place in a democracy, these proponents tend to overlook the benefits that the propertied class sought—conformance, propriety, and ultimately compliance by the multitudes. Critics, on the other hand, focus on the motives of the elites. As we saw, their objection is not to the notion of the public library itself, but the shape that it has been given under elite tutelage. These critics, however, do not provide an alternative vision for the public library if it were to develop without undue elite influence.

When we examine the established order, we see a great deal of coherence—indicative of a generative metaphor guiding the development of the public library's different parts. It is that of a teacher-student relationship. Melvil Dewey himself said that the "librarian is in the highest sense a teacher" (Nardini 2001, 113). Herein, people with knowledge are one side and those deficient in it on the other. The latter come to a noble place—the public library—where the former, invested in uplifting them, attend to their deficiencies. This general point can be elaborated in various ways. For our present analysis, we should note that at the heart of this formulation, there is an asymmetry—one that is unquestioned by those in the establishment. In our analytical vocabulary, the asymmetry is as follows: the system (i.e., the library) is put in place to uplift disadvantaged individuals (i.e., ones deficient in knowledge due to lack of resources).[14]

An asymmetrical relationship is not necessarily bad; it can, in fact, be fruitful. Beyond providing access to books and other information sources, public libraries offer many programs that cater to the local community's needs—digital literacy skills, job search assistance, business skills, health-centered activities, children's recreation and education, career planning for young adults, transition help for veterans returning to civilian life, financial scam prevention workshops for senior citizens, and English conversation groups for immigrants, among others. The public libraries celebrate such accomplishments, and the public seems to be in agreement. In a 2006 telephone survey ($n = 1,203$), the respondents rated public libraries at the top of all services that their local communities do a good job of providing (Public Agenda 2006). In a 2007 online survey of residents ($n = 1,901$) in communities with populations of 200,000 or fewer, 74 percent of the respondents said they would "definitely" or "probably" support funding of the local public library in a referendum, ballot initiative, or bond measure (OCLC 2008). In a follow-up survey a decade later, 58 percent of the respondents said that they would "definitely" or "probably" vote in favor of funding (OCLC and ALA 2018).[15] While the latest survey suggests a softening of support, overall support remains fairly high. Pew Research Center surveys in 2015 ($n=2,004$) and 2016 ($n=1,601$) show that people believe public libraries to be important for their local community, and about two-thirds of the respondents of each survey said that the closure of their local library would have a major impact on their communities (Pew Research Center 2015, 2016).[16]

Given these accomplishments, is it worth our while to look for blind spots in this scheme of things? Our analytical project prompts us to do so.

The public library certainly provides open access—it is available to all, offering its services for free. Furthermore, as noted previously, it does very many good things. One could reasonably argue that this constitutes universal access.

But then, the public library has always, since its inception, had a shortcoming—a significant portion of the population (currently one-fifth, as noted) never uses it. Unlike other systems we have examined, in the case of the public library, it is not difficult to see the problematic at the margins of the established order. Furthermore, here, the establishment has not tried to hide it by obfuscation; on the contrary, it has openly acknowledged the problematic. However, after acknowledging it, the establishment discourse gravitates toward all the things that the public library does well, as opposed to focusing doggedly on this segment of the population. This calls for a recentering-on-reversal.

We know what recentering-on-reversal would entail here: decentering of the beneficent system, and recentering on the one-fifth of the population who never use the public library.[17] Then, we would need to see how things look from their vantage point.[18] But to do all this, we need to know who these people are. Pew surveys provide some idea, but we need a deeper understanding. Without a good understanding of the people who never use the public library, we are limited in what we can do. To get an idea of the kind of fresh thinking that recentering-on-reversal may yield, let us briefly consider the group that Dee Garrison, Oliver Garceau, and other authors have been telling us are disaffected with the public library—workers.[19] If we recenter on workers and see things from their vantage point, we are prompted to consider whether the asymmetrical teacher-student relationship and the attendant tonality are at the root of the problem. Perhaps disadvantaged individuals should be in the driver's seat—not the library. Perhaps the model should be a symmetrical one, wherein the individuals uplift themselves and the library plays only an assistive role. Perhaps the library should be more like a toolshed than a school.

If deeper research into the segment of the population that never visits the library shows that the toolshed model has merit, that does not mean that the existing model, which has its successes, should be discarded. Rather, it means that the existing model needs to be complemented by a toolshed, or some other conception that would work for people who never visit the public library.

7

BROADCASTING

On November 2, 1920, Pittsburgh's KDKA radio station made the famous inaugural broadcast with announcers reading vote counts off telegraph tickers, as they came in, for the presidential election between Warren G. Harding and James M. Cox. It was a watershed moment in the history of broadcasting that has been extensively discussed (e.g., Barnouw 1966; Davis 1930; Douglas 1986, 1987a, 1987b; Hijiya 1992; Sterling and Kittross 1990; Streeter 1996). For us, some of the nitty-gritty is of interest.

KDKA arranged with the *Pittsburgh Post* for telegraph tickers and with federal authorities for broadcast frequencies. It hoped to cover a 200-mile radius. At the last moment, KDKA learned that local radio hobbyists (amateurs) also had plans to broadcast the election results. To minimize conflict and also capitalize on complementary strengths, it forged a collaborative relationship with them (*New York Times* 1924b).

KDKA and the amateurs' American Radio Relay League coordinated to forestall interference on election night—the amateurs refrained from transmitting on a 550 wavelength. Also, they organized community gatherings on election night, where they picked up the KDKA broadcast on their receiver sets and shared them in real time with the attendees over loudspeakers (Urban 1920a, 1920b, 1920c). The joint effort surpassed all expectations.

The returns by wireless telephone, which were transmitted from the Westing-house international radio station at East Pittsburgh, were exceptionally clear and distinct. The service was utilized by many amateurs to entertain gatherings at their various stations. Between announcements of the returns radiophone music was transmitted, which added much to the entertainment. (Urban 1920a, 3)

Here, we see something rare—coexistence of order and un-order.[1] Moreover, this was coexistence in a symbiotic relationship, which is quite extraordinary. Typically, order and un-order tend to be deeply antagonistic and oppositional configurations (Sawhney 2009). Order seeks to organize un-order in its image or stamp it out, and for its part, un-order seeks to break down order—that is, each seeks the elimination of the other. We saw this dynamic play out in epic proportions on the edges of the Roman Empire. For centuries, the expanding empire brought order to the so-called barbarians or kept them on the run. But, when the Romans weakened, the barbarians overran the empire and sacked Rome itself. In this context, the symbiotic coexistence of KDKA (order) and the amateurs (un-order) is worth meditating upon. We will first see how the relationship actually unfolded and then return to discussing the topic at hand, universal access to broadcasting.

EXPANSION OF BROADCASTING

In discussions of universal access, the history of broadcasting is rarely evoked,[2] perhaps because radio (and later television) spread rapidly—KDKA[3] went on the air in 1920, and by 1950, over 95 percent of US households had radio. This growth is visually striking when plotted on a graph alongside telephony, where the corresponding time span is about a century. Moreover, the graph shows uninterrupted growth during the Great Depression (1929–1939). In the post–World War II world, we saw a similar phenomenon with television, but at an accelerated pace—reaching 95 percent of households in two decades.

In the following discussion, we mostly focus on radio because its development greatly shaped that of television, including the organization of the industry and the regulatory framework. In a nutshell, the template for broadcasting was developed with radio and later transferred to television. We will discuss issues specific to television when we examine how notions of universal access developed for radio continue to be operative in the case of television. We will not be discussing other mass media, such as newspapers and magazines (which are covered in chapter 2, on the postal system) or movies, which have a very different history and no tradition of universal access.

INEQUITIES AND INEQUITIES THAT MATTER

On the transmission side of the system, the number of radio stations grew rapidly, but the high national numbers were mostly a product of radio's rapid growth in the cities of the Northeast, North Central, and West regions.[4] The poorer South and rural areas in general had slower growth.[5]

On the reception end of the system, similar regional and urban-rural disparities arose, as did racial ones. In 1930, when 50 percent of urban and 26.9 percent of US households had radio, the corresponding numbers in the South were only 28.6 percent and 9.2 percent. By 1940, over 90 percent of urban and 80 percent of rural households in all regions had radio except the South, where the corresponding percentages were 78.8 percent and 50.9 percent. Only in 1950 did South's average penetration cross 90 percent, with rural penetration still at 88.8 percent (Craig 2004a).

The disparities along racial lines were even deeper. In 1930, when the national average for households with radio was 40.3 percent, the average for those with a Black head of household was 7.5 percent (and only 2.2 percent in the South). In 1940, the census included Blacks in the "nonwhite" category. Since 96 percent of nonwhites were Black, the data provide a good approximation of Black households with radio—only 43.3 percent (and only 16.8 percent of rural nonwhites in the South) had radio, when the national average was 82.8 percent (Craig 2004a).[6]

These disparities were a function of access to two elements: broadcast signals and a receiver set. If one were not within the coverage area of a broadcast station, access to a receiver set would be a moot point. If one were within the coverage area of a broadcasting station, the issue then was whether one could afford a receiver set or access someone else's. With regard to the first situation, we previously briefly noted the regional disparities in the locations of broadcast stations, and in the next section, we will examine this issue at greater depth. For now, though, we will focus on the second situation.

In 1930, when the per-capita income was $1,970, a radio cost $78 on average.[7] Hobbyists and the poor often made their own radios with a "Quaker Oats box, copper wire, and a three-dollar crystal"[8] (Vaillant 2002a, 62). In rural areas, households often had to install long, outdoor antennas to catch distant signals. Since, in 1930, only 10 percent of farm households had electricity (and 85 percent of households in urban areas), farm households, unless they had their own generators, had to rely on battery-powered radios (Craig 2004a; Kirkpatrick 2006).

Over time, the radio's average cost dropped from $38 in 1940 to $26 in 1950. Also, the quality of the hardware and batteries, as well as the availability of electricity, dramatically improved. By 1940, 69.6 percent of rural households had radios (national average 82.8 percent), and by 1950, the penetration was 92.7 percent (national average 95.7 percent) (Craig 2004a). The discourse on inequities typically centers on unserved individuals, as we see with other systems. However, in the case of radio, interestingly, no one called for subsidies or other policy intervention to universalize radios—perhaps due to a set of convergent factors. As discussed previously, the cost of receiver sets was declining, hardware and battery technologies were advancing, and electricity was spreading. In addition, disadvantaged people were showing a remarkable willingness to somehow marshal their resources to purchase a radio, giving it priority over other needs even during the Great Depression.[9] As Barfield (1996) observes, "Whether in the city or the country, the household's first radio was typically borne home like a proud trophy, a symbol of victory in the family budget wars" (15). The public valued news, given the Depression-era politics, and were attracted to entertainment programming (Craig 2009b).[10] Furthermore, the rapid growth of radio, reaching 95 percent penetration within three decades, eased the buildup of public pressure for a policy intervention.[11]

Interestingly, the inequities that mattered were areal ones—regional and urban-rural. Southern elites feared domination by Northern broadcasters, and the rural elites by urban broadcasters—displacement of particularistic localism by universalistic cosmopolitism was the issue at heart. Consequently, the universal access discourse in broadcasting has centered on localism.

THE CALL FOR LOCALISM

In broadcasting policy, localism is a signature American issue, and until recently an exclusively American one.[12] Even so, despite persistent deliberation, it remains an ambiguous notion, as we will see next.

In 1926, the US Congress passed legislation to create the Federal Radio Commission (FRC) for the management of spectrum assignments. The next year, it enacted the Radio Act of 1927,[13] the legal basis for FRC's localism policy: "In considering applications for licenses and renewals of licenses . . . the licensing authority shall make such a distribution of licenses, bands of frequency of wave lengths, periods of time for operation, and of power among different States and communities as to give fair, efficient, and equitable radio service to each of the same" (Section 9).

The commitment to localism was deepened by the 1928 amendment (Davis Amendment) to the Radio Act, which read as follows: "people of all the zones . . . are entitled to equality of radio broadcasting service, both of transmission and reception, . . . licensing authority shall as nearly as possible make and maintain an equal allocation of broadcasting licenses, of bands of frequency or wave lengths, of periods of time of operation, and of station power, to each of said zones."[14]

On the surface, this orientation toward the local in broadcasting is a recent manifestation of a deep-seated American cultural imperative to disperse power.[15] In reality, it was the product of a political fight. When Congress passed legislation to create the FRC in 1926, it authorized one-year tenure for the new agency, with the possibility of reauthorization. Once the FRC started operating in 1927, its decisions raised concerns among legislators, especially from the South, who felt it was favoring large broadcasting corporations, especially those based in New York and Chicago.[16] To enforce a corrective, Representative Ewin Davis (R) of Tennessee added the Davis Amendment to the reauthorization bill, saying: "We want broadcast licenses fairly distributed in such a manner that those who desire to do so may listen to New York and chain stations when they want to, but may, when they so desire, listen to broadcast by stations elsewhere throughout the country, including their own zones, states, and cities" (Kirkpatrick 2006, 284).

In this conception, the local licensee was the "bedrock of the broadcast system"—seen as vital for acquiring programs tuned to local needs, on the one hand, and servicing as a platform for local expression by producing local programs, on the other (Horwitz 1989, 157).

But that did not come to pass, for two primary reasons: consensus on the definition of localism proved elusive, and the rise of networks profoundly changed broadcasting.

ELUSIVE DEFINITION OF LOCALISM

When the FRC started formulating its localism policy after the passage of the Davis Amendment, the difficulty of defining localism immediately came to the fore:

There was disagreement among the commissioners as to the precise meaning of the amendment . . . immediate reallocation of the broadcast band . . . a policy to be followed in the future . . . wherever possible . . . the question of whether the amendment required an equality of the number of licensed stations without regard to division of time or whether two or more stations dividing time could be balanced against one full-time station in another zone (Messere 2005, 44).

Broadly, the FRC could reduce inequity by decreasing the number of stations in some areas and increasing it in others; the former created political problems because communities wanted to keep their stations, and the latter created spectrum interference problems. Further, beyond the number of stations, the FRC also had to consider wattage, as it had a bearing on a station's reach, as well as the potential for interstation interference. Furthermore, the FRC had to consider proportionality with population, since the Davis Amendment called for equitable allocations "according to population." In the face of these complexities, the FRC came up with a tiered system—national, regional, and local stations, which was controversial from the outset (Messere 2005, 43).[17]

In subsequent decisions by the FRC and the Federal Communications Commission (FCC)[18] (e.g., licensing, spectrum allocation, ownership rules, and must-carry regulations), localism continued to loom big, but in the same diffused way as described previously (Horwitz 1989). Napoli (2001) provides an extensive review of the different definitions of localism used in different decisions. Broadly, they tend to center on the notion that broadcasters need to be responsive to the geographic communities for which they were awarded a license (Ali 2017).

THE RISE OF BROADCASTING NETWORKS

Within a decade of the passage of the Radio Act of 1927, with the development of broadcasting networks, the broadcasting landscape changed in ways that the lawmakers had not anticipated.[19] They had focused on the physical location of the transmitter, assuming that it would foster a close link with the nearby community, and in turn that would shape the programming.

With the development of networks, local stations became local affiliates, and the network–local affiliate relationship was an open question—the possibilities ranged from a loose federation to a tight integration. In a loose federation, the local affiliates would be primarily locally oriented entities that occasionally came together for national broadcasts of special events. In a tightly integrated system, networks would provision programming centrally and the local affiliates would serve as delivery points. The former would have preserved local autonomy, but the latter had seductive aspects.[20]

The early years of the affiliate-network relationship were marked with these tensions—conflicts and negotiations over local versus centralized control. Over time, the seduction offered by the near-continuous flow of network-provisioned programming proved too tempting—it removed the affiliates' burdens of producing expensive programming or procuring it from different suppliers, and it also lightened the burden of securing advertisers because the network handled much of that duty centrally (Horwitz 1989, Kirkpatrick 2006, Socolow 2001, Streeter 1996). From the national system perspective, a tightly integrated system served another purpose—it provided a restraining hold on the local.

The case of William Kennon Henderson's high-powered station KWHK, broadcasting from Shreveport, Louisiana, to large parts of the South and the Midwest, is telling. In his "bare-knuckled" broadcasts, Henderson launched tirades against the FRC, broadcast networks, chain stores, Wall Street, and other extraregional institutions, on the one hand, and a staunch defense of Southern culture (using rather coarse language) on the other. The FRC sought a way to shut down KWHK for years. In 1933, when an opportunity finally arose to shut it down, FRC offered Henderson a deal: if he affiliated KWHK with CBS, he would be allowed to retain his broadcasting license. Henderson refused, and KWHK was taken off the air (Vaillant 2004, 193). Here, we should note that the FRC saw network affiliation as a means of restraining the idiosyncratic local—and, interestingly, so did Henderson.

Conversely, the predominance of metropole-centric integrated networks stifled the development of the local public sphere, the ideal lying at the heart of localism.[21] The relationship between radio and the local public sphere before and after the rise of networks begs for interrogation, which we will now undertake.

NETWORKS AND THE LOCAL PUBLIC SPHERE

In 1921, only a handful of experimental stations, including KDKA, were broadcasting. Within three years, the US had over 500 licensed radio stations (Craig 2004b).[22] About 30 percent of them were nonprofits (run by colleges, religious groups, labor unions, civic associations, and others) broadcasting locally produced programs. By the end of the decade, almost all the nonprofits were off the air (Streeter 1996). In this brief period, "millions of ordinary people, individually or in local organizations, produced countless hours of local civic, educational, religious, and social programming on local stations, usually on a sustaining basis" (Kirkpatrick 2006, 18). While the nonprofits were unable to survive in a world remade by the FRC and broadcast networks, voices for diversity persisted, especially those of educators.[23]

For its part, the FRC also purportedly sought to develop the local public sphere with its localism policies. As we saw, its (and later the FCC's) localism efforts were transmitter-centric, based on the assumption that a local transmitter would catalyze locally originated programming. With the advent of television, localism again became a concern, and here too the focus was on the location and power of transmitters, but now the FCC also made a conscious effort to encourage locally originated programming. Over the years, the FCC's localism efforts took many different shapes, such as making local program development a criterion for license renewal, requiring broadcasters to create local production facilities. However, none of them lived up to expectations, and they were eliminated; today, all that broadcasters are required to do is to provide a local or toll-free telephone number (Braman 2007; Feldman 2017; Napoli 2001).[24]

With the availability of portable and cheaper cameras and recording equipment, new possibilities for nonexpert programming started opening up; for instance, the Public Broadcasting Service (PBS) began airing materials produced by local nonprofit groups, with its technical assistance. The advent of cable television relaxed the constraints on the distribution side, opening up more new opportunities for nonexpert programming. Now, entire channels could be dedicated to bottom-up programming, as wired cable networks had the capacity to carry many more channels than just the regular broadcast channels. The community media activists clamored for such access,[25] and the cable companies were willing to oblige, as it ingratiated them with local governments (Engelman 1990).

Interestingly, this confluence of interests of community media activists and cable companies led to something remarkable—a capitalist-funded platform (public access TV centers and access channels) for playing out the radical impulses of community media activists (Klein 2006). These centers—more than 2,000 of them, with most towns and cities having one—provide access to production equipment, training, and technical assistance, and that too was on a nondiscriminatory first-come, first-serve basis. The result has also been remarkable—they have had limited success, at best.

Klein (2006, 8) provides an incisive assessment, which takes into account "four institutions" that structure public access television: cable technology, regulatory framework, organization design, and professional culture. In Klein's pithy characterization, public access television centers have basically become video clubs. While the users of the centers' resources ostensibly do so to produce programs for the general public, the viewers have little presence in the centers—either physically or in the discourse of the producers.[26] Moreover, beyond the walls of the centers, the general public has little awareness about the programs produced there, let alone viewership, and consequently they make at best a limited contribution to the creation of the local public sphere.

After many repeated efforts at developing localism, one could say that it is now something merely symbolic. As Braman (2007) points out, the reality is that "local broadcasting transmission stations are no longer stand-alone institutions in the way that the local printing presses were during the colonial period but are merely the localized faces of the global infrastructure" (240). The net result is that much of programming continues to come from outside and contributes little to the creation of a local public sphere (Braman 2007). What is striking is that the discourse of localism continues to persist. Regulations ask for localism and are justified on the basis of localism. Conversely, arguments and petitions are made in the name of localism and on the basis of localism. No matter how con- voluted the logics become, the discourse persists. Cole and Murck (2007) called it the "myth of the localism mandate." This persistence points to a tension—loss of the particular in the world of the universal.[27]

At the same time, we should not idealize the local—mythologizing it into an idyllic past—because it has its own peculiar warts. When network broad- casting started, the industry made a strong distinction between "indirect" advertising, which was okay, and "direct" advertising, which was a taboo (St. Austell 1928, 58).[28] The former took various forms (e.g., company-owned radio stations run for publicity and goodwill, or company-sponsored pro- grams on commercial stations). The notion of indirect advertising was elas- tic, and advertisers pushed the limits, but the existence of limits weighed on their minds. Edgar Felix, a leading advertising expert, called for subtle writing because the "radio audience . . . resents the slightest attempt at direct advertising" (Smulyan 1994, 76). As an example, he gave the case of the Happiness Candy Company, which "got over the idea that their stores are conveniently located throughout New York, without resorting to a direct advertising statement, by working in the phrase that 'happiness is just around the corner from you'" (Smulyan 1994, 76). Eventually, the line was breached—but not by networks.

In 1927, a small, rural station—KFNF in Shenandoah, Iowa—broke the taboo against direct advertising with Henry Field's[29] program, where "the public was begged to send orders for tires, dishes, peaches, coffee, Chi- nese baskets, pencils, fountain pens (guaranteed for life) (sic) suits, over- coats, paint, canned corn and nursery stock, not forgetting prunes" (St. Austell 1928, 58). KFNK fans were ardent followers of Field and vocally defended him in public forums.[30] Nationally, there was an uproar. The FRC considered banning direct advertising but decided that such an action was unnecessary, expecting it to die on its own in the face of lis- tener pushback, given the offense it caused (Craig 2010a).

Thus, ironically, advertising-supported network broadcasting, which undermined localism, was birthed by a taboo-breaking local radio station in rural America. Moreover, in the pre-FRC period, localism was not even an issue—broadcasts by amateurs and nonprofits were essentially expressions of localism. In this light, one can only wonder about regulators' faltering efforts to foster localism—something they quashed when it existed in a very organic form.

GAINS AND TRAVAILS

We could have developed our broadcasting system in various ways—the system we have was not inevitable, just a possibility among others. The scholarship on sociotechnical systems has shown this to be true of systems in general, so broadcasting is not special in this regard. However, it is special in that its materiality is less of a constraining factor than that of other systems, since transmission towers are easily deployable, and in many different ways. In effect, we had many more possibilities with broadcasting, and yet we ended up with the one we did, which begs for a close examination of the gains and travails of the individual and the system in this realm (table 7.1).

Table 7.1 Universal access: Gains and travails

	Gains	Travails
Individual	Integration with global circuits of information flows—news, crop prices, weather reports, etc Access to professionally produced entertainment of a wide range Reduction in rural isolation	Encroachment of particularistic local culture by universalistic cosmopolitan culture of the metropoles Erosion of locally organized cultural activities whose value primarily lay not in the final product but the process of creation itself, which helped build communal bonds
System	Tighter integration of rural areas into the metropolitan economy Expanded markets for corporations Tutoring of consumption on an industrial scale	Initial difficulties in setting up the system for regulation of broadcasting

INDIVIDUAL

Gains Country dwellers could keep abreast of developments in the world beyond their own immediate environs, something that they particularly valued during the Depression, when distant events were upending their lives (Craig 2009b, 1). They were particularly interested in programs such as NBC's *National Farm* and *Home Hour*, which informed them of New Deal programs for rural America[31] (Craig 2009a, 2009b). They also gained from access to quotidian but critical information, such as weather reports and crop prices (Gardner 2009; Wik 1988). In the words of Kenyon Mix, an agriculture researcher: "The city man listens with mild amusement to an announcer's recitation of a long list of prices on hogs, corn, wheat, butter, eggs, cream and potatoes, but the farmer listens with deep concern, for it affects his personal welfare" (*New York Times* 1926, 8). In general, surveys showed that rural listeners placed news on the top of the benefits of radio (Craig 2009b; Gardner 2009; *New York Times* 1926; Wik 1988).

Rural families typically placed the radio—a "secular altar" of sorts—in a central place in the house, where they would gather and listen together (Jellison 1993, 94). In the mid-1920s, about 90 percent of radio programming was music-based; such shows were cheap to produce and popular with audiences (Craig 2009b). In particular, *WLS National Barn Dance*, a precursor to the *Grand Ole Opry*, was extremely popular.[32]

Compared to Rural Free Delivery (RFD) and telephony, both seen as reducing rural isolation, radio was much more reliable, as it continued working during the toughest days of winter. Moreover, it helped reduce isolation in new and more immediate ways (Kline 2000).

Rural families were avid listeners, with 90 percent listening every day and for an average of about 5 hours each day, approximately 15 percent more than their urban counterparts (Craig 2010b). We find vivid expression of these statistics in a 1938 contest organized by *Rural Radio* magazine on the theme "What radio means to my family" All twelve winning entries, showcased in the May 1938 issue, dwelled on either the reduction of isolation or the benefits of connection, such as news, entertainment, and knowledge (*Rural Radio* 1938).

Travails The elites tended to see rural problems as stemming from rural "deficiencies" (i.e., shortfalls vis-à-vis urban areas). For instance, the report of the President's Commission on Country Life in 1909 identified "unequal development of our contemporary civilization" as the cause of rural-urban migration (Craig 2009a, 13)[33]. Correspondingly, the elites sought to eliminate this deficiency—that is, extend the "contemporary civilization" (urban) to areas deficient in it (rural).

Ideally, for such a project, broadcasting would take national programs of "contemporary civilization" to all people, a capacity that it did have—technologically. The reality, however, turned out to be much more complex. Such programs often had limited appeal in rural areas, and moreover, tradition-minded people resisted them (e.g., dismissing all new forms of music as "jazz"). They preferred "old timey" music rooted in traditional music forms (Craig 2009a; Vaillant 2002b). Similarly, programming, especially jokes, that touched on culturally sensitive issues such as race and sexuality alienated country dwellers (Kirkpatrick 2006). More broadly, they did not like the tone of such programs, sensing it to be condescending (Vaillant 2002b). Network executives, on the other hand, thought that "the non-cosmopolitan audiences were stubbornly, mysteriously clinging to their cultural tastes and preferences" (Kirkpatrick 2006, 282).

But then, for them, country dwellers were also an unsatisfied market. Like good marketers, the networks came up with a formulation that worked for this audience and was profitable for themselves to boot. They developed programming with a local feel, celebrating simplicity, neighborliness, and independence.[34] For instance, in *The Real Folks of Thompkins Corners*, people did little more than good, neighborly things—something that real people in real small towns across America could relate to (Kirkpatrick 2006). But then, "Thompkins Corners" was a generic small town. As Barron (1997) explains, in such programs, the networks basically "institutionalized localistic values of homeliness and neighborliness in ways that transcended the particular community, and it helped to define a more general culture that celebrated localism without being directly tied to the culture of any one locality" (225). This formulation, which Barron aptly calls "translocal" popular culture, attracted audiences from particularistic small towns, ironically, with a universalistic product created on an industrial scale (222).[35]

Those of us who watched television in the 1980s and early 1990s partook of the trans-local formulation in the hugely popular situation comedy *Cheers*, set in an eponymous bar "where everybody knows your name." It offered the feel of a familiar spot in your neighborhood—but it was not. It could be in any neighborhood. Moreover, it felt like a place where everybody knew each other's names—but in reality "everybody" did not know our names (i.e., real people with real names). It was a synthetic offering, a therapeutic one.

While *Cheers* offered a synthetic balm of sorts for the artifice of modern life, trans-local broadcasting took a toll on rural communities, which still had the real thing—a place where everybody *actually* knows your name. It eroded, if not destroyed, local entertainment, especially the Chautauqua meetings, named after Chautauqua Lake on whose shore the first meeting was held in 1874. In these meetings, which lasted from three days to a week, local communities put together programs mixing sermons, self-improvement lectures, music, yodeling, cartooning, and other entertainment by local talent and invited outsiders. Later, even when communities started outsourcing programming to Chautauqua circuits (traveling companies), they still had to do considerable local organizing—picnics, games, costume parties, and so on. It was anything but a sterile affair, as a traveling Chautauqua lecturer tells us:

A more heterogeneous audience you could not find than the crowd which sits under the average brown tent. The ninety-year-olds are there and the nine-month-olds. There is no Chautauqua circuit in this country which would succeed if you barred babies from the tents. For Chautauqua is essentially a family institution, and the farmer's wife cannot leave the house unless she takes her baby with her. (Mason 1921, 418)

President Theodore Roosevelt called the Chautauqua meetings "the most American thing in America." (Canning 2005, 227). The commercial Chautauqua circuits sapped the Chautauqua spirit, and radio pretty much destroyed it; the former provided prepackaged solutions that were convenient and of professional quality, and the latter took these seductions to an entirely different plane.

Radio also undermined the local churches. In 1931, Orestes Caldwell, a commissioner of the FRC, said of radio and the local church: "In place of the choirs, church leaders may bring to their fellow communicants . . . the life like voices of metropolitan vocal stars . . . Even the preacher himself may have to give way to noted divines whose sermons, carried far beyond metropolitan pulpits, will be audible to countless thousands of Sabbath worshippers in 'electronic chapels' in villages, towns, and tiniest cross-road hamlets" (Kirkpatrick 2006, 18). In its obviousness, certainly to Commissioner Caldwell, such thinking misses something critical— the process of creation itself is often more important than what actually gets produced. It builds communal bonds. As Kirkpatrick (2006) notes, "the local church choir is not solely about musical quality, but also about participation in local public spheres" (18). Similarly, a sermon, even by "noted divines," that speaks to everybody in a universalistic trans-local way lacks the particularistic resonances that brings singularly cosituated peoples into a fellowship.[36]

The trans-local has order and tidiness. But it is also sterile. It is telling that "when it came to perhaps the touchiest subject of all, religion, NBC quickly learned not to work with local churches or individual religious figures, producing all of their religious broadcasts as sustaining programs at the national level" (Kirkpatrick 2006, 274). On a more subtle level, the network announcers were directed to speak in "hushed tones, emphasizing clear diction, avoiding an identifiable regional accent, and exuding a calm demeanor" (Kirkpatrick 2006, 281). In a pithier formulation, *Radio Broadcast* magazine pronounced that the best announcer is "one who is only slightly more human than an automaton" (Kirkpatrick 2006, 281).

SYSTEM

Gains   Nord (1986) argues that "radio was not the first mass medium for the live 'broadcast' of the human voice. The railroad was. The railroad made possible a new profession for America in the nineteenth century: lecturing" (16). Speakers, making a living giving lectures, delivered the same orations thousands of times across the country. For example, Russell Conwell, a famous Baptist preacher, supposedly gave his "Acres of Diamonds" sermon over 6,000 times. Similarly, the railroads carried music, vaudeville, and other shows across the country. Quite clearly, society, as it was industrializing, had a need for the circulation of uniform voice messages across space, prior to the arrival of radio (Nord 1986). Radio served this need much more efficiently. It also helped resolve the "crisis of control" created by industrial production outstripping consumer demand.

By raising consumption, broadcast advertising aligned consumption with production (Beniger 1986, 6). For such an alignment, rural areas were critical; until 1920, the majority of the US population lived there.[37] In Dyer-Witheford's more political formulation: "the market requires a consuming subject, a subject that needs what capital produces and believes that these needs can and must be satisfied in commodity form" (1999, 117). To the corporations, broadcasting provided a vehicle for tutoring consumption on an industrial scale.

Travails Both the corporations and the government were blindsided by the emergence of broadcasting. They focused on the development of radio as an extension of the telegraph network, an addendum that expanded its reach to places the wires could not connect (e.g., ships at sea). Accordingly, they sought to develop radio as a point-to-point technology, an orientation announced in its byname—wireless telegraph. Holding such a conception, they saw radio waves' tendency to spread as a nuisance, and also a security risk—something they needed to tamp down. On the other hand, amateurs, transmitting from their garages, exalted in radio waves' tendency to spread, and for them, the so-called security risk was an opportunity to develop an audience (e.g., transmit music for their fans, who in turn sent postcards with thanks, requests, and so on). In effect, the amateurs, not the corporations or the government, identified a new mode of communication: point-to-multipoint (aka broadcasting).

When broadcasting, at first a peripheral activity, entered the mainstream, it upended the established frameworks, including General Electric (GE)-RCA-Westinghouse-AT&T patent pooling agreements, which were based on a point-to-point understanding of radio (Brock 1981). Also, the courts were at a loss, as thus far, they had been judging radio cases on the basis of telegraph precedents (Sawhney, Suri, and Lee 2010). Moreover, the government faced a new kind of governance problem—interference—and it had to figure out a solution quickly, even when the spectrum was not well understood.

In sum, the birth of broadcasting was a shock to the system. On the other hand, once a system was developed to regulate broadcasting, the general template continued to serve over the years, including for television later, with relatively small modifications along the way.

DECENTERING INTERFERENCE

The initial impulse for regulating broadcasting—removal of interference—developed into a totalizing imperative to bring all of broadcasting into a system, a unitary national one. To fashion such a system, the powers that be brought broadcasting under federal authority via the Commerce Clause (Article 1, Section 8) of the US Constitution, which grants the Congress the power "To regulate Commerce with foreign Nations, and among the several States, and with the Indian Tribes" (Constitution of the United States 1788). This required two conceptual moves.

One was to establish the truism that broadcasting is inherently a commercial activity. There is no doubt that some broadcasting is commercial activity, but the notion that all of it is so is easily refutable—just consider various forms of nonprofit broadcasting. Yet such a case was made. It had to be made, as it took care of the "commerce" requirement of the Commerce Clause. Its backers offered a whole slew of arguments in support of such a formulation, none of which was persuasive on its own accord. Let us look at two from a comprehensive review provided by Kirkpatrick (2011): radio was basically wireless telegraph, and since all telegraph is commerce, so should broadcasting be; after all, even nonprofit broadcasting has a financial dimension. Furthermore, the court rulings that established broadcasting as commercial activity did not spell out their rationale in a definitive way.

The second was to establish the truism that broadcasting is inherently an interstate activity. There is no doubt that some broadcasting is an interstate commercial activity, but the notion that all of it is so is highly questionable—just consider a low-powered transmitter with 1-mile range in the middle of Texas. Yet such a case was made. It had to be made, as broadcasting had to be not only a commercial activity but also an interstate one—"among the several States." Here, its backers made the argument that even a low-powered transmitter could cause interference with an interstate broadcast. Theoretically, it is true. But *could* does not mean *will*, even in theory. More important, the courts bought this argument (Kirkpatrick 2011).

It all came together in the *Whitehurst v. Grimes* (1927) decision of the US District Court (Eastern District of Kentucky), overturning a local tax on an FRC-licensed amateur radio broadcaster:

Radio communications are all interstate. This is so, though they may be intended only for intrastate transmission; and interstate transmission of such communications may be seriously affected by communications intended only for intrastate transmission. Such communications admit of and require a uniform system of regulation and control throughout the United States, and Congress has covered the field by appropriate legislation. It follows that the ordinance is void, as a regulation of interstate commerce.

Here, in a 292-word decision that also includes introductory case details, the court accepted that radio broadcasting is interstate commerce, without seeing any need to explain whatsoever. This is very remarkable, given what a critical decision this was.

One can accept the argument that we needed a system. But it need not have been the "uniform system" that the court talks about. There is nothing axiomatic about such a system—in reality, many types of systems are possible in any domain. We can get a sense of the possibilities just by relaxing the two purported constraints that cast broadcasting as interstate commerce.

If we free ourselves of the notion/constraint that broadcasting is inherently commercial, right away our mind space is populated with two categories of broadcasting: broadcasting that is commercial and broadcasting that is not.

If we free ourselves of the notion/constraint that broadcasting is inherently an interstate activity, right away our mind space is populated with two other categories of broadcasting: broadcasting that is interstate and broadcasting that is intrastate.

With these categories, we have a matrix of possibilities at a very basic level: commercial-interstate, commercial-intrastate, noncommercial-interstate, and noncommercial-intrastate. No doubt, commercial-interstate would come under federal jurisdiction and commercial-intrastate and noncommercial-intrastate under states' jurisdiction. On the other hand, noncommercial-interstate would not be such a clear-cut case, and therefore it will be the focus of our discussion.

In the US, we do have experience with joint federal-state regulatory regimes (e.g., telephony). However, in the case of noncommercial-interstate broadcasting, we might not be able to neatly demarcate regulatory jurisdictions. But then, we do not have to rule out less-than-perfect solutions, or even make-do ones. We could develop innovative joint federal-state mechanisms. In its day, the creation of the Tennessee Valley Authority called for many joint federal-state innovations. Shortly, we will see how Germany uses collaborative mechanisms between states to deal with interstate issues in broadcasting. The bigger point here is that any such multidimensional system would be much more complex and messier than a unitary federal one, which is not necessarily a bad thing.

Ostensibly, the problem of interference made the formation of a unitary federal system unavoidable. On the other hand, the powers that be pursued a zero-tolerance policy toward interference with a zeal that begs for interrogation. Further, this policy produced a system that served them all too well. All in all, it is very likely that the zero-tolerance policy was a convenient proxy for their goals. If this had not been the case, it is very unlikely that they would have pursued it with such zeal that it became totalizing.[38]

If we decenter interference as the central organizing issue, we start seeing new possibilities. Clearly, any broadcasting system would need to remove interference. But does it need to be removed entirely? What if the focus is not on total elimination of inference, but on keeping it below a specified threshold? For instance, we could tolerate some interference in the night, when radio waves propagate farther, while having clear broadcasts during the day.[39] It will be a messier situation, but we could still have a viable broadcasting system.

Nobody asked this question, as far as we can tell. Many people criticized myriad aspects of the FRC's regulatory regime, but its interference policy was not made an issue. Our analysis suggests that that was a mistake. Tolerance of the possibility of some interference, even in the abstract, would have relaxed the hold of the logic behind a unitary system, if not unraveled it altogether.

As noted earlier, the reality is that systems need not be unitary. Moreover, it is possible to have no interference in a nonunitary system. Take the case of Germany, where both the federal and state (*Länder*) governments regulate broadcasting. The sixteen *Länders* have their own broadcasting laws on matters related to the organizational structure and financing of public broadcasters, as well as the licensing of private broadcasters. When interstate issues arise, they forge interstate agreements, sometimes involving all the *Länders* and at other times only neighboring ones. "These inter-state instruments have a similarity with bilateral or multilateral agreements under international law. They are drafted by state governments concerned and subject to ratification by their respective parliaments" (German Law Archive n.d.). The federal government has sole regulatory authority over technical matters related to the transmission of broadcasts and copyright. The Federal Constitutional Court rules on interstate and state-federal disputes and ensures that the fundamental rights enshrined in the Basic Law are protected (German Law Archive n.d.). Here, in sharp contrast to the American system, we see inbuilt complexity to the levels seen in international agreements.

Ironically, the US has a rich and deliberate tradition of complex governance arrangements, with myriad mechanisms for checks and balances. Yet, when it came to broadcasting, the powers that be made the case that it is so singularly exceptional that its governance had to be unitary. To add to the irony, the complexly decentralized German system was deliberately developed by the US and the other Allies during their post–World War II occupation of Germany, to diffuse control and forestall the rise of another authoritarian government[40] (Reich 1963; Toogood 1978).

Now, with a braced openness to messy solutions, let us reconsider the FRC's (and later the FCC's) priorities, which would normally not merit a second thought because they are reasonable and normal for our times.

One, the FRC instituted high technical standards for granting licenses. Who can argue against high technical standards? But the reality is that the FRC's high technical standards favored the commercial broadcasters and worked against the noncommercial ones, as the former could afford expensive equipment and the latter could not (Messere 2005). Many small operators could not keep up with the FRC's increasing technical requirements, a task made even more difficult when the FRC also started assigning them the least desirable parts of the spectrum, and they were forced to shut down (Streeter 1996). Moreover, in 1928, the FRC's General Order 40 for nationwide channel allocation skewed things in favor of high-power stations—forty high-power national clear channels, thirty-four high-power regional channels, and thirty low-power local channels in each of the five zones. As McChesney (1993, 26) notes, after creating these high-powered national and regional channels, "the FRC argued that it was obviously in the public interest to assign these channels to broadcasters who had the equipment to take advantage of these slots." In effect, the FRC thereby set up a technical justification for the bias toward well-capitalized entities—and conversely, a bias against marginal ones.[41]

Two, the FRC privileged programming with high production values, introducing a bias toward network broadcasting. *Radio Broadcast*, a trade magazine, opined that "it can be said without fear or favor that chain broadcasting is responsible almost entirely for the growth of high-grade programs in this country" (1928, 65). On the other hand, the emphasis on high production values worked against amateur productions. For instance, the number of higher education–affiliated stations declined between 1925 and 1930 from 128 to 42, principally because of lack of funds.[42] By 1934, all of nonprofit broadcasting amounted to only 2 percent of total broadcast time in the US (McChesney 1993).

Three, the FRC called for comprehensive programming or a "well-rounded program," which served the entire public in the licensed area (McChesney 1993, 27). Who could argue against comprehensive programming? But the reality is that there is an inbuilt bias toward cosmopolitanism in such an approach. The local is often parochial, something pejorative in our times. The FRC called stations of special interest groups (e.g., educators, religious groups) "propaganda" (28) stations, in marked contrast to the broad-based "general public service" (28) stations. In its *Third Annual Report,* the FRC stated outright that a general public service station had "a claim of preference over a propaganda station" (McChesney 1993, 28). This insistence on comprehensiveness not only eliminated idiosyncratic voices, but it also increased the seepage of the cosmopolitan culture into the parochial. While broadcasters developed specialty programs for rural audiences, often of the trans-local variety, their efforts were primarily directed at comprehensive programming, consequently, once a rural community came on the broadcasting system, its exposure to the cosmopolitan urban culture increased significantly. Moreover, given the economic incentives, broadcasters standardized comprehensive programing across the country, creating a trans-local bundle of sorts. In effect, the FRC's insistence on comprehensive programming ironically worked against diversity (see Vaillant 2002b for vivid examples of how the rich diversity of Chicago broadcasting was wiped out).

Four, FRC purportedly treated all the stations the same. What could be wrong with that? Its standards were cued to those of commercial broadcasters. In effect, though, the FRC's uniform approach had an built-in bias against nonprofit broadcasters. In the words of Harold Lafont, a serving FRC commissioner, in 1931: "under existing law the Commission cannot favor an educational institution. It must be treated like any citizen, any other group, any other applicant. I see nothing in the law that would justify the Commission's doing otherwise, regardless of our interest in education" (McChesney 1993, 34). If diversity of voices is indeed important, the FRC's approach here is highly questionable.

Five, the FRC thought in terms of a broadcasting system. Accordingly, the focus was on nationwide optimization—alignment of all the pieces into a unitary whole, such that *aggregate* output is maximal. In such a framework, "channel assignments are justified in terms of the expected outcome for the entire system of broadcasting, not in terms of intrinsic values as freedom and fairness" (Steeter 1996, 100). For instance, in its first move to reduce interference,[43] the FRC sought to remove 164 marginal stations; when it stopped halfway through because of political backlash, that required the surviving stations to lower power or share frequencies (McChesney 1993; Messere 2005). More bluntly stated, the FRC's focus was not on local flourishing. On the contrary, it sought to remove or tamp down local idiosyncrasies and untidiness—quirks that give character to a place.

In sum, the reality is that the systems, born of the priorities and motivations of our time, are biased against localism—even more so when there is an insistence on unitary or uniform systems. At their core, our systems are trans-local projects. We deploy them to coordinate across space, increase efficiencies, and expand economics of scale. In such formations, the connections among places are more important than the places themselves, save the centers of power. As such, the seductions of system solutions are very high.

As Kirkpatrick (2006) tells us, even before the rise of broadcasting, "rural and small-town newspapers had long used syndicated material to construct a generic 'local' identity for themselves, running nationally-distributed cartoons and columns proclaiming the virtues of 'your local community' or 'your hometown'" (290). So much so that rural and small-town newspapers even used syndicated materials for "buy local" campaigns (Kirkpatrick 2006). Today, we are seeing such system solutions seep into all areas of life; for instance, even restaurants that emphasize their local, nonchain essence use precut, packaged, and semiprocessed materials. Against such a backdrop, localism policies are at best remedial actions that are likely to have limited impact. There is no escaping this. What we learn from the present analysis is that the chances of attaining some degree of meaningful localism are likely to be higher in messier systems than in totalizing ones.

Returning to the opening discussion on the symbiotic coexistence of order (KDKA) and un-order (amateurs) at the time of the Harding-Cox election broadcast, we need to reflect on the path taken and alternatives forgone. Was it right of policymakers to allow, if not aid, order to stamp out un-order? Or should they have countered order's efforts to stamp out un-order, preserving at least some of it? These were real options. In the years preceding the passage of the Communications Act of 1934, the Payne Fund and its allies pressed Congress to require the FRC by law to set aside 15 percent of channels for nonprofit broadcasters. This could have plausibly happened if the politics had been different. In fact, two months before the 1934 law was enacted, the Wagner-Hatfield amendment, reserving 25 percent of channels for nonprofits, had a good chance of adoption (McChesney 1993). Likewise, policymakers could have given amateurs space to flourish instead of deliberately marginalizing them (Sawhney and Lee 2005). They were a source of innovation, including the development of shortwave radio technologies, and could have remained so.[44] In effect, coexistence of order and un-order was possible, at least on some level.

The policymakers had an urban-centric view, seeing radio as a vehicle for extending the "contemporary civilization" (urban) to areas deficient in it (rural). In a rural-centric view, the priority would be the other way around: to "put an urbanizing society back in touch with its rural, small-town roots" by inducing "cities to listen to the hinterlands" (Kirkpatrick 2006, 121). In such an effort, we would focus on preserving local idiosyncrasies as opposed to leveling them out.

Ideally, we would have cultivated a balanced blend of both—urban cultures enriching rural ones and vice versa. What we have actually produced, though, is a denuding asymmetry of near-total urban domination. But then, realistically, even if we had worked to achieve a balance, we would still have asymmetry—albeit not so severe. On the other hand, with policy-makers' overriding focus on developing a unitary system, asymmetry was almost certain, and an extreme one at that. A messier system, with coexistence of order and un-order at some level, would have allowed something less skewed, as well as some possibility of meaningful localism, to exist.

8

THE INTERNET

Hush-A-Phone, a funnel-shaped telephone attachment invented in 1921, guided the air propagating the user's voice into the receiver. Three decades later, with more than 100,000 pieces sold, AT&T forbade its use. When Hush-A-Phone Corporation filed a complaint with the Federal Communications Commission (FCC), it ruled in AT&T's favor, identifying the plastic contraption as an "unauthorized foreign attachment" (Brock 1981; Zittrain 2008). However, in 1956, the DC Circuit Court of Appeals ruled against the FCC, opening up the telecommunications arena for innovation.

This court ruling later guided the FCC's 1968 decision on Carterfone, another telephone accessory of sorts—a coupling device that facilitated conversation between one party on a telephone and the other on a two-way radio, making the connection acoustically rather than electrically. Yet AT&T prohibited its use on the ground that the acoustic connection infringed on its system. The FCC ruled against AT&T and decided to allow not only Carterfone, but also electrically connected devices, so long as they did not cause harm to the telephone system, opening the gates for innovations such as the answering machine, the fax machine, and the cordless phone. The great, transformational innovation was the dial-up modem, which enabled computer networking over telephone lines. The rest is history (Brock 1981; Zittrain 2008).

EXPANSION OF THE INTERNET

Stalwarts of the Internet Society[1] mark their telling of the internet's history with milestones such as the following:

- Leonard Kleinrock's publication, in 1961, of his first paper on packet switching[2]
- J. C. R. Licklider's articulation, in 1962, of his vision of the Galactic Network, a worldwide network of interconnected computers
- The deployment of the Defense Advanced Research Projects Agency (DARPA), in 1969, of the first links of the ARPANET
- The National Science Foundation's funding of a national backbone network for the internet, in the 1980s, and the subsequent expansion of access to universities, research institutions, and the private sector
- The Clinton-Gore administration's commercialization, in the 1990s, of the internet,[3] and its global expansion, in the 2000s (Leiner et al. 1997)

In a similar vein, Thomas Hughes describes the development of the internet as "a memorable, salient example of the manner in which ARPA, using a light touch, funded and managed the rapid development of high-risk-high-payoff computer projects" (Hughes 1998, 255). He talks of a high-tech culture—nonhierarchical, meritocratic, and multidisciplinary—with proclivities for modularity, interconnectivity, and transformational change. In highlighting ARPA's "light-touch" approach, Hughes emphasizes the role of "managerial prowess," which, he believes, largely explains American dominance in the high-tech realm (Hughes 1998, 5).

In actuality, such accounts are limited, since the internet was a product of a much broader social churn. As Flichy (2007) tells us, "When we leave 'the short-term dimension' of technical development, that is, a specific project, and consider a more long-term dimension such as electrical light and power, high-speed trains, internet, and so forth, we encounter more than simply a project or common intention; what we witness is a collective vision or *imaginaire*" (4, italics in original). In other words, the internet *imaginaire* extended beyond project managers, academics, government officials, and entrepreneurs. For instance, Hu (2015) shows how artists and activists contributed to the conceptualization and development of early networks. Among other things, *Radical Software*, a newsletter of a group of activist engineers, carried network designs eerily similar to ARPANET's, and the

Truckstop Network—"a countrywide network of truck stops for 'media nomads'"—anticipated today's participatory culture (28). More broadly, Hu (2015) argues that the internet was not just a creation of DARPA, but the product of a time when "the center was not holding," as Joan Didion (1990, 84) famously noted.

Whatever version of history one holds, it is undeniable that the biggest turning point was the commercialization of the internet. It took the internet beyond the rarified world of defense research agencies, universities, and research institutions and thrust it into the tumult of the marketplace. The internet spread across the US and the world beyond, with all manner of entrepreneurs, hackers, and activists opening up new ways of doing things. In short, commercialization radically changed the scale and scope of the internet.

INEQUITIES AND INEQUITIES THAT MATTER

Broadly, our understanding of "access" has undergone three major changes over the years, which are as follows.

EXPANSION OF THE INTERNET: ACCESS AS PHYSICAL CONNECTIVITY

Until 1994, the US Census Bureau and the FCC collected data on telephone penetration, but not on computers and modems. Then, as the National Telecommunications and Information Administration (NTIA) explains in the introductory discussion in its first *Falling Through the Net* report, it realized: "While a standard telephone line can be an individual's pathway to the riches of the Information Age, a personal computer and modem are rapidly becoming the keys to the vault" (NTIA 1995, 2). Subsequently, the NTIA paid the Census Bureau for including computer and modem ownership questions in its thrice-yearly Current Population Survey. Talking about the "have nots" and "information disadvantaged," based on the survey data, the report showed that these people were disproportionately in rural areas, inner cities, and minority populations. In general, it also indicated that the less educated were less likely to have telephones, computers, and modems. The second NTIA *Falling Through the Net* report, issued in 1999, talked in terms of the "digital divide" (NTIA 1999).

The *Falling Through the Net* reports set the proverbial table for digital divide research. Subsequently, researchers studied many dimensions of the digital divide, especially race, gender, urban versus rural, age, education, and income. Studies on race and the digital divide showed that minorities were on its wrong side (e.g., Fairlie 2014; Hoffman and Novak 1998; Ono and Zavodny 2002). In the case of gender, researchers found that while the gap persisted, it was narrowing (Dholakia 2006; Losh 2003). On the other hand, in the case of urban-rural disparity, the researchers found that the gap persisted, and it was not likely to narrow (Strover 2001, 2003). Studies on the elderly also showed a divide and spotlighted barriers to internet use (see Xie 2003 for a review). With regard to the last two factors, education and income, researchers examined them as part of analyses of a specific population or comprehensive studies of the digital divide (e.g., Bucy 2000; Lenhart and Horrigan 2003; Robinson, DiMaggio, and Hargittai 2003). In general, researchers painstakingly recorded inequities that gave rise to the digital divide. Moreover, they were of one mind that these inequities mattered. The dispute was on what to do about them—specifically, the need and justification for public policy intervention.

Dismissing calls for policy intervention as "ideological manipulation," the opponents argued that over time, markets universalize products and services as technological advances, economies of scale, and competition bring down prices. In effect, they say that the so-called digital divide is a transitory phenomenon (Compaine 1986, 2001b; see Lentz 2000 for a survey of skeptics). In this vein, the former FCC chairman Michael Powell derided calls for policy intervention by likening digital divide to a "Mercedes divide" (Irving 2001, A30). Fundamentally, the opponents' views are undergirded by theorists such as Daniel Bell, who believed that "technology has not only raised the standard of living but it has been the chief mechanism of reducing inequality within Western Society" (Compaine 1986, 8). To substantiate their argument, opponents point to actual experience with technologies such as radio, television, and automobiles.[4] They drew two lessons from this experience:

(1) Policymakers should be mindful of the long arc of technology development, and not succumb to short-term, fashionable pressures. Given enough time,

markets bring prices down to levels within everybody's reach. In short, the policymakers should refrain from intervening prematurely.

(2) When the policymakers do intervene, they should basically "fine tune." The reality is that development dynamics great vary across technologies. The universalization of automobiles did not require a policy intervention, telephone's initial expansion was a product of competition and only later regulatory interventions were made to universalize it, broadcasting did not require direct subsidies to users and the policy interventions were programming related. In short, a blanket approach for universal access is a folly (Compaine 1986).

In sum, while everybody agreed that these inequities mattered, there were two schools of thought on what should be done about them. One advocated policy intervention, and the other called on policymakers to refrain from intervening in the markets since they universalize technologies over time. Both sides, as we will see, were partly correct.

EXPANSION OF THE INTERNET: ACCESS AS PARTICIPATION

With the internet's rapid expansion in the 2000s, its proponents' concerns about physical connectivity were alleviated. Then their attention moved to knowledge and skills that people need to use the internet, benefit from online resources, and participate in online forums. They started arguing that physical connectivity by itself does not ensure inclusion and participation (Hargittai 2002; Loader and Keeble 2004; Newhagen and Bucy 2004; Servon 2002; Stevenson 2009; van Dijk 2005). At this point, the universal access discourse shifted from connectivity to inclusion and participation.

To justify the resources required, the proponents broadly highlighted two benefits. One, digital skills enhance democratic participation, a long-running concern given declining civic engagement and voting turnout. Two, they enable disadvantaged individuals to secure jobs and also, correspondingly, expand the labor pool for an information technology (IT)–intensive economy. In actuality, the economy also needed digitally skilled consumers, with e-commerce on the rise.

Skeptics of technology-based solutions to social problems pointed to America's long history of "ascriptive hierarchy—social exclusion on the basis of foreordained characteristics such as race, gender, and ethnicity (Mossberger, Tolbert, and McNeal 2008, 7). They said that the digital divide is basically a reproduction of long-entrenched inequities in the online domain (Margolis and Resnick 2000; Xenos and Moy 2007). Yet, ultimately, even they ended up calling for policy intervention for universal access because it advances educational opportunity, which they believed was critical for leveling the economic playing field (Mossberger, Tolbert, and McNeal 2008).

In reality, as we will see, the internet's expansion is a product of a far more complex logic than the exclusionary logic of ascriptive hierarchies or the inclusionary logic of equity-oriented digital divide policies.

EXPANSION OF THE INTERNET: ACCESS AS CONTRIBUTION

Broadly, in the pre-internet era, a rather clear division of labor existed in the production, transmission, and consumption of media content: producers (film and TV studios, writers, journalists, and musicians, among others) developed creative output, delivery channels (film distributors, broadcasters, and cable systems, among others) distributed it, and the general public consumed it. Today, in our world of smartphones and internet uploads, the entire process is scrambled. Now, for the first time in human history, ordinary people can produce content and also reach mass audiences at very low cost—enabling crowdsourcing (Shirky 2008), sharing economy (Benkler 2006), and co-operativism (Conaty and Bollier 2014).

Even in this low threshold condition, we find echoes of the digital divide. Researchers report gender divide in contributions to Wikipedia (Cohen 2011; Hargittai and Shaw 2015). In the case of crowdsourced online cartographic resources like OpenStreetMap and Google Map Maker[5], researchers report gender divide and also urban-rural divide in participation and contributions (Hecht and Stephens 2016; Johnson 2016; Stephens 2013). More broadly, based on data from seventeen Pew Internet & American Life Project national surveys taken between 2000 and 2008, Schradie (2011, 151) found that while user-generated content of all types has grown rapidly, the poor and the working class have created such content at a slower rate than rest of the population—creating a "production divide."

To gain a fuller understanding of the digital divide, we must also consider the other side of user-generated content production. The digital environment supposedly operates according to the logic that "the more we put in the more we get out"—our summation of Mark Zuckerberg's point of view (Zuckerberg 2012). Ekbia (2016), following Terranova (2000), counteracts with the logic: "The network gets from you much more than you get from the network" (172). The profits generated by social media companies are evidence of this—such profits are totally dependent on the presence of people on their networks and the content they create. In effect, universal access expands the pool of people whose data and content social media companies can harvest.

We hear echoes of these discourses, albeit at times faintly, in the chapters of this book on other systems. This is not surprising, as the internet is protean, complex, and multidimensional. Physical connectivity was also an issue in the case of the postal system, electricity, and telephony. Participation was also an issue in the case of education and public libraries. Contribution, however, was an issue only in the case of broadcasting, though at a different level. While universal access efforts for broadcasting promote diversity of sources of content production at the regional and organization levels, the issue for internet policies is diversity at the level of the individual. Yet the experience with broadcasting prompts us to pause and reflect on our approach for the internet. For instance, in the literature, researchers talk of the development of the digital divide in terms of levels—first-level, second-level, and third-level (e.g., Hargittai 2002; Min 2010; Scheerder, van Deursen, and van Dijk 2017). However, as we see in chapter 7, on broadcasting, which is an outlier, issues related to content production are very different than those related to physical connectivity and participation. This suggests that we are misguided in treating contribution as the "next" (i.e., third) level of digital divide and, more generally, thinking of universal internet access in terms of levels.

GAINS AND TRAVAILS

We will now take stock of the gains and travails for the individual and the system (table 8.1).

INDIVIDUAL

Gains The COVID-19 pandemic brought forth the benefits of internet access to the individual in a singular manner. In 2019, about 7 percent of workers in the private industry, mostly white-collar professionals, had

Table 8.1 Internet: Gains and travails

	Gains	Travails
Individual	Fuller participation in the social, economic, and democratic processes of a technologically advanced society Opportunities to reduce costs for transportation, entertainment, and services Opportunities to engage in new modalities of sharing, as well as creating and disseminating content	Cost of equipment and service Hazards of online fraud Loss of privacy; marginalized losing one of their few advantages—diffused visibility Social media activities creating a record Blurring of public-private boundary
System	The state more efficiently delivering services and also surveilling Private companies leveraging an infrastructure created with public funding to make profits Funding of a digital divide "industry"—equipment and service providers, funding agencies, nonprofits, consultancies, academic and nonacademic researchers	Institutions, both public and private, more vulnerable to hackers The power grid and other infrastructures more vulnerable to hackers Greater possibility of foreign interference in domestic affairs

opportunities for telework. In April 2020, soon after the start of the pandemic, about half of the US workforce was working from home (DeSilver 2020; Guyot and Sawhill 2020). Only individuals with internet access could do so (and it had to be of adequate quality). Individuals living in areas without stable broadband access had to go to their workplaces in order to have Zoom meetings with their colleagues, who were now working from home (Merrefield 2020).

We saw something similar with college students, who were suddenly asked to vacate campuses. While on campus, all students had access to the internet, but upon moving back home, many did not have access to stable broadband, or even any kind of broadband, for participation in Zoom class sessions. In the world of K–12 schooling, we even have a term for such inequities—the "homework gap," which became a bigger problem during the pandemic (Ali 2020). In another critical area of life—health care—telehealth assumed a new importance during the pandemic. About 50 percent of doctors were using telehealth in April 2020, as opposed to about 18 percent in 2018 (Landi 2020). In effect, individuals needed internet access to participate in critical activities and access critical services.

Even when access is available through offline channels, online access is very often advantageous. Like other airlines, for instance, United Airlines levies no ticketing fee for online purchases, but it charges for purchases by phone ($25) and in person ($30 at a city ticket office, and $50 at the airport).[6] Furthermore, while purchasing online, one can avail oneself of many more options and access special deals. In a similar vein, while shopping in a brick-and-mortar store, it is worth one's while to check the price for the *same* item on the website of the *same* company—often the price on the website is lower, and, moreover, the brick-and-mortar store will often match it (Gerdeman 2018). Marketing literature even has a term for it—"self-matching" (Kireyev, Kumar, and Ofek 2017). Furthermore, online comments help one make better purchasing decisions, even when in a brick-and-mortar store. The tag on the shoes says "Waterproof"—but is it? Online comments of consumers who have already worn the shoes can provide critical insight on this score. Finally, businesses offer discounts, which can be significant, for online quotes, online signing, paperless billing, and other electronic interactions.

Internet access enables individuals to participate in the sharing economy. We have been seeing the emergence of informal sharing communities that exchange occasionally used items, ranging from tools to party supplies, with each other. Some of them grow into formally organized cooperatives, often with membership fees, such as "tool libraries" (Shmurak 2016). Individuals are also able to participate in online cooperative activities—writing and editing Wikipedia pages, developing open-source software, clicking on images of an asteroid to help the National Aeronautics and Space Administration (NASA) map its surface, among others (Bartels 2019). Of course, individuals can also participate in various online forums on topics ranging from pop culture to politics.

Travails To access the internet, the individual has to pay for connectivity. In 1999, the average household expenditure for internet access was $49 per year (Bureau of Labor Statistics 2012). By 2017, the average cost of a broadband plan was $66.17 per *month* (McCarthy 2017). In 2017, house-holds also spent an average of $94 per month on cellular phone service, another means of accessing the internet (Bureau of Labor Statistics 2019). In addition, the individual pays for devices (computers, tablets, Wi-Fi routers, etc.) and premium services (Netflix, Pandora, Spotify, etc.).

Beyond the direct costs of internet access, the individual is exposed to the hazards of phishing, identity theft, and other types of online fraud. In 2019, the Internet Crime Complaint Center of the Federal Bureau of Investigation (FBI) received 467,361 complaints, reporting losses totaling over $3.5 billion. The top five reported complaints were related to phishing/vishing/smishing/pharming, nonpayment/nondelivery, extortion, personal data breaches, and spoofing (FBI 2020). These statistics do not include online tax fraud. In 2016, cybercriminals attempted to siphon off at least $12 billion in fraudulent tax returns and got away with at least $1.6 billion (Davidson 2018).

Another gauge of online fraud is the size of the identity theft protection services industry—there are 135 companies offering such services, with a total revenue of $2 billion (*IBISWorld* 2019). Furthermore, businesses like Facebook and Google are now providing identity services, intruding into a domain that had been the exclusive preserve of government agencies that furnish passports, driver's licenses, and other such documents (Morozov 2014).

The individual also suffers loss of privacy, as her online activities are logged by internet service providers, companies, and others. For the marginalized, the loss is doubly severe since they also lose one of the few advantages that their marginality affords—diffused visibility. They are now registered with all manner of systems and their activities logged by companies and organizations with online tracking technologies. Without media literacy, they are further disadvantaged, as they lack the awareness and knowhow to look for errors and obtain corrections in databases.

More generally, once online, the individual's life is marked by pressures that blur the boundary between public and private realms. This blurring of the public-private boundary predates the internet. Carey (1989) laments the rise of the Sunday newspaper in the late nineteenth century, which eroded the Sabbath—"a region free from the control of the state and commerce" (227). The rise of radio took the blurring of the public and private domains to a new level, as strangers from the realm of commerce could now "talk" to people, including children, in their private spaces. The internet has taken it to yet another level—we are now to varying degrees 24/7 consumers and 24/7 workers.[7]

SYSTEM

Gains The state gains from efficiencies in administration. For instance, it costs the Internal Revenue Service (IRS) only $0.35 to process an e-filed tax return, as opposed to $2.87 for a paper return. Furthermore, e-filed returns have an error rate of less than 2.5 percent, as opposed to over 25 percent for paper returns (Mock 2015). Similarly, in the case of food stamps, which literally had been stamped paper or coupons, the state moved to electronic benefit transfer (EBT) in the 1990s, motivated by opportunities to cut administration costs and fraud. In the early years of EBT implementation, the situation with regard to administration costs was mixed, with them decreasing in some states and increasing in others, but over time, they have been reduced. EBT has also reduced fraud (Isaacs 2008; Stegman, Lobenhofer, and Quinterno 2003). Beyond efficiencies in administration, the state's capacity for surveillance has also increased with the enmeshment of the internet in everyday life (Braman 2006; ProPublica 2013; Zetter 2016).

Businesses gain from internet-enabled lower transaction costs in their everyday processes, from placement of orders to invoicing, as well as from outsourcing of processes to specialized service providers (for extensive reviews, see van Alstyne 1997 and Swanson 2020). They are able to collect consumer data at an unprecedented scale and harvest user-generated content for monetization in various forms, including videos on YouTube, travelers' comments on TripAdvisor, and personal news and memories on Facebook. Furthermore, they are able to capitalize on new configuration potentialities—new modes of connectivity that previously were not possible at scale—that are afforded by the internet in order to develop new types of business such as eBay, Uber, and Airbnb (Sawhney and Lee 2005). In these ways, businesses have capitalized on the capacities of an infrastructure created with public funding.

Universal access involves billions of dollars. For instance, in 2019, the FCC's Universal Service Fund disbursed $8.3 billion. The Universal

Service Administrative Company, the nonprofit company that administers the Universal Service Fund, itself has 570 full-time employees (Universal Service Administrative Company 2020). The recipients of these monies—telecommunications service providers, schools, libraries, rural hospitals, and others—have their own employees for universal access issues. Furthermore, contractors, equipment suppliers, consultants, and others are also involved. Academic research on the digital divide itself is a cottage industry of sorts. All in all, universal access itself has become an industry (Sawhney 2003, Stevenson 2009).

Travails Beyond public relations disasters, data breaches also cost businesses financially—over $8 million on average. Small businesses are the most frequently affected (constituting 60 percent of victims), and the hospitals are the worst affected (Ponemon Institute and IBM Security 2020; Puranik 2019; Walker 2019). The cost of cyber/data breach insurance provides another window into the travails of data breaches. For instance, in rounded numbers, a hospital pays a $250,000 annual premium for $10 million worth of coverage, a university foundation pays $17,000 for $2 million of coverage, and a vacation rental pays $1,000 for $1 million of coverage (Marciano 2020).

The electricity distribution system's vulnerability to hacking became salient when Ukraine's grid suffered an attack in 2015. For the US, protection against such a cyberattack is particularly challenging for a number of reasons. First, its power system is a complex assemblage of 3,300 utilities working together to provide electricity over thousands of miles of transmission and distribution lines. Second, the US is a highly networked nation, with millions of smart devices that are potential gateways for determined attackers. Third, it relies on privately owned electric utilities, which shy away from making investments in cybersecurity technologies given their business model (Knake 2017; Smith and Barry 2019). More broadly, private corporations play a major role in the US, owning about 85 percent of critical infrastructure assets, and the government has had difficulties in incentivizing them to invest in cybersecurity (Etzioni 2011; Gordon, Loeb, Lucyshyn, and Zhou 2015).

The possibility of foreign interference in domestic affairs has greatly increased, both in intensity and scope. For instance, the roots of Russian disinformation go back to the Soviet Union. Joseph Stalin himself coined the term *dezinformatsiya*—strategy directed at disorganizing in stealth an opponent's information capability. Soviets sought to somehow fool the mainstream media into disseminating disinformation, starting with an obscure news outlet and then working up to major ones. In a particularly notable case, in the early 1980s, a small pro-Soviet newspaper in India carried an article saying that the Pentagon in the US had created the AIDS virus. Subsequently, citing this article, a Soviet weekly published its own story on this, and eventually a British tabloid ran the story on its front page. Thereafter, major newspapers in over fifty countries picked up the story (Bailey 1987; Taylor 2016).

In other words, the recent Russian interference in the American political process is not novel. The impulse has been around for a while—but what has changed is that the internet has, on the one hand, allowed its intensification, and, on the other, provided direct access to individual Americans. In today's media ecology, RT (formerly Russia Today) TV provides fodder for Twitter, Facebook, and other social media stories and in turn serves as an outlet for spurious so-called fake news bubbling up in social media, and every now and then, this process widens to loop in other news outlets, often major ones.

CONNECTIVITY AND ALSO ITS NATURE

The internet is now woven into the fabric of society, and, consequently, it has become enmeshed in the everyday life of the individual. For the individual, the draws of the internet are clear and alluring, but its pitfalls, while generally known, lack clarity, given the complexity and opacity of the commercial relations that undergird its myriad services.[8] In our analytical terms, the individual's experience of the gains is vivid and the individual's understanding of the travails vague. In such a context, it is difficult for the individual to assess the situation, weigh the trade-offs, and navigate the gains and travails of internet access.[9]

On social media, one reconnects with old classmates and starts exchanging messages, posting old photos, and cracking jokes. A classmate posts an old photo, which one wishes she had not. Soon an individual starts seeing regular snippets of his life broadcast, often for all to see. While individuals are generally aware that their postings are curated presentations of self to others in an arena that allows a high level of control, one forgets and slips into envy, which is real, even if the imagery that provoked it is not. Further, a person can become very conscious of the "likes," "follows," and other social affirmations that posts receive or do not receive. In addition, there are advertisements, increasingly tailored ones, and prompts to connect with someone who has "502 friends." Individuals are constantly tempted into the tangle of this weave, promising amusement, humor, voyeurism, reactivation of memories, and other benefits. On disconcerting occasions, when an individual considers "getting off the grid," the costs weigh on the mind—such a person will be left out of social get-togethers, among other things. While all this is true of sociality in general, on the internet there is an intensification of its allures and demands.

While ordinary people are caught in this web, loosely aware of the travails, the elites are calibrating their own engagement in a deliberate manner. Ironically, the very people responsible for social media place limits on their children's engagement with them. Steve Jobs did not let his children use iPads; Bill Gates gave his children cell phones only when they turned 14; Evan Williams, cofounder of Twitter, gave his children hard-copy books in lieu of electronic tablets; Evan Spiegel, the chief executive officer of Snapchat, limited his children to 1.5 hours of screen time per week (Akhtar and Ward 2020; Bilton 2014; Bradshaw 2018; Rudgard 2018). In fact, Chris Anderson, the chief executive officer of 3D Robotics, talking about his five children, aged six to seventeen, said:

My kids accuse me and my wife of being fascists and overly concerned about tech, and they say that none of their friends have the same rules. That's because we have seen the dangers of technology firsthand. I've seen it in myself, I don't want to see that happen to my kids (Bilton 2014, n.p.).

Furthermore, they engage in resistance, digital retreat, disconnection, and media abstention. They have summer camps, literally, for themselves. For instance, Digital Detox LLC offers Camp Grounded: Summer Camps for Adults, characterized on its website as follows:

Campers from over 30 states and 8+ countries have traded in their computers, cell phones, emails, Instagrams, clocks, schedules, work-jargon and networking for an off-the-grid weekend of pure unadulterated fun. Together we create a community where money and titles are worth little . . . and individuality, self-expression, friendship, memories and the great outdoors are valued most. Ages range from 18–75 . . . offers 50+ Playshops & Activities . . . Counselors and more. Just like the summer camp you remember from childhood, but for adults.[10]

Interestingly, in the discourse on universal access, disconnection and nonuse are seen as a "problem" that needs to be "solved," an abnormality, a pathology—"a deficit on the part of the individual concerned" (Selwyn 2003, 106). Ironically, in their embrace of resistance, the elites have, at least for themselves, recast disconnection and nonuse as a virtue. But what about the populations served by universal access programs?

Universal access programs focus on connectivity. Through the evolution of universal access programs, from physical connection and participation to content contribution, policymakers and researchers have taken internet connectivity to be inherently good. But we have seen in this discussion that connectivity is good and bad—complexly so.

Certainly, universal access programs should continue to provide connectivity. The internet has become a consumption norm, which Preston and Flynn (2000) explain to be the bare essentials required by the poorest citizen to function effectively in society. Revisiting the seminal writings of Adam Smith, they show that consumption norms are a function of the wealth and sociomaterial characteristics of a society, as opposed to some absolute standard. For instance, even the rich in ancient Greece did not wear shoes, something essential for even poor workers in our society. In modern society, the internet is a consumption norm. Indeed, even the elites who are committed to resistance and disconnection do not give up connectivity altogether—they do so only for short periods.

Similarly, instead of simply taking connectivity to be inherently good, universal access programs need to also factor in the complexities of connectivity. In our analytical terms, not only the gains but also the travails of connectivity should be considered. Nonusers, especially those who can afford connectivity, are the difficult parts of the intellectual system that informs universal access. We have been explaining them away as pathological. Instead of discounting them, if we center-stage these nonusers and see the world from their standpoint, we start to see the complexities of connectivity.

The reality is that connection has its problems, and so does disconnection. Our blind spot has been that in our preoccupation with connectivity, we have not considered the question: What should be the *nature* of connectivity?[11]

We will engage with this question in some depth in the next chapter. At this point, to register its significance, we revisit two commonplaces in the domain of universal access. One, we routinely use frequency of use as a metric for measuring the digital divide.[12] Is high frequency of use a good thing? Should policies be directed at increasing frequency of use? What about the travails of high frequency of use? On the design front, systems are designed and celebrated for making the internet frictionless. Is that a good thing, or should we insert deliberate speed bumps that prompt moments of pause before we make decisions/nondecisions in our world of instant gratification?

9

CONCLUSION

In this final chapter, we have organized the discussion in two parts: summative and reflective. In the summative part—the first two sections—we bring together key findings from system-specific chapters into overarching frames. In the reflective part—the last two sections—we take a final look at the project as a whole and delineate its big takeaways.

In the first section, we focus on the gains and travails identified in the system-specific chapters. We provide comprehensive tables for individual's gains, individual's travails, system's gains, and system's travails, which draw together the gains and travails from the tables in the system-specific chapters. In these tables, we also note whether each specific gain and travail was discoursed.[1] In the accompanying commentary, we focus on gains and travails that were either not discoursed at all or not fully discoursed.

In the second section, we focus on blind spots and the utility of the recentering-on-reversal strategy in various contexts. We have created a comprehensive table that brings together findings from the system-specific chapters on the nature of each blind spot identified, the reasons for its occurrence, the utility of recentering-on-reversal strategy in the context it occurred, and our takeaway for future analysis of other systems. In the accompanying commentary, we synthesize the individual findings.

In the last two sections, we offer our final reflections on the project and discuss its potential contribution to universal access debates beyond the American shores.

GAINS AND TRAVAILS

The tables provided in this section are essentially quadrants of the gains and travails typology, as shown in figure 9.1.

In these tables, for each gain and travail, in the last column, we note whether it was discoursed. We use the following notations: yes, mostly, partly, and no. If a gain or travail was discoursed, we do not discuss it in greater depth, as we seek to take the discourse beyond what is said to what is *not* said. Accordingly, in the following discussion, we focus only on the gains and travails that were not discoursed at all or not fully discoursed (i.e., those with the notations "no," "partly," or "mostly").

INDIVIDUAL'S GAINS

As we see in table 9.1, across all seven systems, almost all the gains to the individual were discussed. We find only two exceptions, both of which were partly discussed: education's social benefits and public libraries' aid to the homeless. In the case of the former, the social benefits of education were extensively discussed, but subsequent gains, such as women's empowerment, were not considered. Similarly, in the case of the latter, the social benefits of the library were extensively discussed, but specific future gains, such as aid to the homeless, were not part of the discourse. Both speak to the nature of such systems. On the one hand, they create

	Gains	Travails
Individual	Table 9.1	Table 9.2
System	Table 9.3	Table 9.4

9.1 Gains and travails, quadrants and tables.

Table 9.1 Individual's gains

	Gains to the individual	Discoursed?
Postal system	Convenience of home delivery	Yes
	Increased access to crop prices, weather reports, educational and other information resources	Yes
	Improved road connectivity	Yes
	Increased value of land	Yes
Education	Increased earning power	Yes
	Skills to function in the world beyond one's immediate environment	Yes
	Social benefits such as improved sanitary practices, family planning, and women's empowerment	Mostly
Electrification	Improved quality of life	Yes
	Increased productivity on farms	Yes
Telephony	Reduced barriers of distance for social communications	Yes
	Reduced barriers of distance for business communications	Yes
	Enables calls for emergency assistance	Yes
Public libraries	Free access to books and other information sources	Yes
	Librarians often going the extra mile to provide information on government agencies and community organizations that could be of help	Yes
	Libraries as communal spaces	Yes
	Libraries helping the homeless	No
Broadcasting	Integration with global circuits of information flows—news, crop prices, weather reports, etc.	Yes
	Access to professionally produced entertainment of a wide range	Yes
	Reduction in rural isolation	Yes
Internet	Fuller participation in the social, economic, and democratic processes of a technologically advanced society	Yes
	Opportunities to reduce costs for transportation, entertainment, and services	Yes
	Opportunities to engage in new modalities of sharing, as well as creating and disseminating content	Yes

capabilities that the powers that be desire, motivating them to marshal societal resources for their development. On the other hand, once created, such systems' capabilities can be used for other purposes by the establishment as well as other actors, often in subversive ways, including by those on the establishment's payroll.

INDIVIDUAL'S TRAVAILS

In table 9.2, we see that travails to the individual were not fully discussed in the following cases: the postal system, electrification, telephony, and the internet. We consider the first three of these systems together and the fourth one separately, since they bring into play different issues, as we will see later in this chapter.

Subtle Recontextualization In the cases of the postal system, electrification, and telephony, the nonconsidered travails were a product of a subtle recontextualization of the individual's everyday life—over time, incremental and gradual changes amassed into a totalizing transformation. As Rural Free Delivery (RFD) expanded, the postal service secured address changes on one rural road, and then another, and eventually a nationwide, comprehensive, grid-based addressing system was in place, refashioning the local environments of rural communities, erasing local idiosyncrasies, and enabling extralocal flows. Electrification eased the labor of farming, vegetable and fruit processing, canning, and such undertakings on one farm, and then another, and another, but the cumulatively increased farm productivity also reduced the need for labor, prompting more migration to cities. Telephony first connected farmers to their neighbors, and then to businesses in nearby towns, and eventually to the world at large, engendering asymmetrical, extralocal relationships that eroded local autonomy. All these changes were subtle and gradual and very consequential in the long term, as they recontextualized the individual's rural life—ironically so, as the development of these systems was motivated by their proponents' desire to preserve rural communities.

Discounted Travails In the case of the internet, much has been said about online fraud, loss of privacy, traces and records created by one's social media activities, and the blurring of the public-private boundary. However, from

Table 9.2 Individual's travails

	Travails to the Individual	Discoursed?
Postal system	Exposure of local small businesses to competition from urban-based corporations	Yes
	System's imperative to install standardized mailboxes, assume snow removal responsibilities, etc.	Yes
	Refashioned local environment with the imposition of a grid-based addressing system	No
Education	State's encroachment on parental control	Yes
	Secularized education that does not offend anybody	Yes
	Centralization and bureaucratization of education	Yes
	Prioritization of needs of state and economy over idiosyncratic flourishing of children	Yes
Electrification	Increased productivity on farms, reducing the need for labor, prompting more migration to urban areas	No
Telephony	Dividends of connectivity are dependent on power differentials between the parties connected	No
	Consequences of scaling-up of telephony from area networks to a global one were not understood; scaled-up networks engender asymmetrical extralocal relationships	
Public Libraries	Part of the array of disciplinary institutions that seek to mold the individual	Yes
	Alienating environment in regard to class sensibilities	Yes
	Having to pay taxes whether or not one considers the library to be a public good	Yes
Broadcasting	Encroachment of particularistic local culture by universalistic cosmopolitan culture of the metropoles	Yes
	Erosion of locally organized cultural activities whose value primarily lay not in the final product but the process of creation itself, which helped build communal bonds	Yes
Internet	Cost of equipment and service	Yes
	Hazards of online fraud	Partly
	Loss of privacy; marginalized losing one of their few advantages—diffused visibility	Partly
	Social media activities creating a record	Partly
	Blurring of public-private boundary	Partly

our standpoint, it is important to note that these travails are absent in universal access discourse, which has focused on the costs of connectivity, and lately also on the skills needed to make effective use of connectivity after it is provisioned. If these travails hold weight for the population at large, why are they discounted for disadvantaged persons receiving subsidized access? It is as if connectivity trumps everything else. Should it? We will revisit this question a little later in this chapter.

SYSTEM'S GAINS

As we see in table 9.3, almost all the gains to the system were discussed, often in a general way. We note a number of them as being mostly discoursed because they were not fully spelled out. For the enhanced security capabilities of the state, as with the postal system and the internet, this is very understandable given the imperatives of security. In other cases, for a complete understanding of universal access, a fuller articulation was called for. For instance, at one point, the Rural Electrification Administration (REA) threatened to go beyond power distribution to power generation in order to secure cooperation from private utilities, which also sold power to rural cooperatives. Clearly, the REA knew that the sale of power was a gain to the private utilities and used it as leverage. Yet it did not fully articulate it as such in rural electrification discourse, which remained skewed toward individual's gains and system's travails. In a similar vein, the discourse on public hygiene could have gone beyond the humanitarian to the economic payoff—cost savings and productivity gains; the discourse on telephony beyond gains to independents from the latent demand that their networks revealed to the entire economy; and the discourse on broadcasting beyond the expanded reach of advertising to tutoring of consumption on an industrial scale to better align production, distribution, and demand. Finally, in the service of the candor that we seek, we should acknowledge that the digital divide is now an industry involving billions of dollars of subsidy flows and millions of dollars of research and implementation grants, as well as thousands of employees of equipment and service providers, funding agencies, nonprofits, consultancies, and academic and nonacademic researchers.

Table 9.3 System's gains

	Gains to the System	Discoursed?
Postal system	Increased sale of newspapers and periodicals	Yes
	Increased reach of advertisers	Yes
	Increased business for mail-order companies	Yes
	Enhanced administrative and security capabilities of the state	Mostly
Education	Educated workforce	Yes
	Increased capacity to create and staff complex systems in industry, military, medicine, etc.	Yes
	Savings such as in healthcare with improved personal hygiene	Mostly
	Assimilation of immigrants	Yes
Electrification	Utilities supplying wholesale power to rural electricity cooperatives	Mostly
	Electrical appliance manufacturers profiting from expansion of markets for their products	Yes
Telephony	Revelation of untapped latent demand in rural areas it had failed to fathom	Mostly
	Quickened pace of technological advance	Mostly
Public libraries	Libraries complementing the school system	Yes
	Libraries providing opportunities for the development of information technology skills, which are important for today's economy	Mostly
	Libraries helping to dispel discontent by reducing ignorance and contain it by providing harmless entertainment	Yes
Broadcasting	Tighter integration of rural areas into the metropolitan economy	Yes
	Expanded markets for corporations	Yes
	Tutoring of consumption on an industrial scale	Mostly
Internet	The state more efficiently delivering services and also surveilling	Mostly
	Private companies leveraging an infrastructure created with public funding to make profits	Mostly
	Funding of a digital divide "industry"— equipment and service providers, funding agencies, nonprofits, consultancies, academic and nonacademic researchers	No

SYSTEM'S TRAVAILS

In table 9.4, we see that all the travails to the system were discussed in all the seven cases, albeit not in our analytical terms.

Tables 9.1 through 9.4 are deliberately constructed artifacts. We have slotted in them gains and travails that we identified in the course of our research, seeking to be as exhaustive as possible. In actuality, though, these gains and travails were not discussed in this manner, let alone in one forum. In these tables, we brought together a forced gathering of the sorts

Table 9.4 System's travails

	Travails to the System	Discoursed?
Postal system	Conflict between postal service and publishers	Yes
Education	Unavoidable trade-offs of the sort that give rise to unhappy constituencies	Yes
	Expansionary tendency—there are always constituencies pushing for upping the level of universal education to higher grades in school, to community college, to four-year college	Yes
Electrification	Tumult generated by strenuous and sustained political pushback by those opposed to government's involvement in electrification	Yes
Telephony	Caught off guard by the emergence of competition in rural areas	Yes
	Access issues attracted federal scrutiny	Yes
Public libraries	Only about half the population visiting a public library at least once a year	Yes
	A sizable section of the population (about one-fifth) never visiting a public library	Yes
Broadcasting	Initial difficulties in setting up the system for regulation of broadcasting	Yes
Internet	The institutions, both public and private, more vulnerable to hackers	Yes
	The power grid and other infrastructures more vulnerable to hackers	Yes
	Greater possibility of foreign interference in domestic affairs	Yes

of expected benefits and concerns that various stakeholders have voiced here and there. On the other hand, for our part, in the future, we hope that they will stimulate conversations that put all the gains and travails on the proverbial table, help us develop a fuller understanding of universal access, and create grounds for formulating and implementing thought-through policies.

BLIND SPOTS AND RECENTERING-ON-REVERSAL

As we have seen in the preceding chapters of this book, recentering-on-reversal is not a sure-fire strategy. On the contrary, it is a strategy for developing exploratory probes that prompt us to go beyond the articulated and tease out the unarticulated. Its payoffs vary (see table 9.5).

In our cases, it was particularly helpful with the individual's travails that are subtle, prompting us to look for them and tease them out, as in the chapters on education and the internet. In the case of education, since the various stakeholders took the social system to be a given, the task of education was reduced to molding the individual to fit in. They explained away the consequent molding pressures (e.g., compulsory attendance laws, time discipline of schedules) as being in the individual's best interest. Looked at another way, though, they are the individual's travails, and, in this light, they energize the question: should children be molded to enhance the well-being of society, or should society change to enhance the flourishing of children? In the case of the internet, connectivity is taken to be inherently good, and people who willfully choose not to connect are taken to be an oddity or our "difficult bit." They are pathologized—something is wrong with them. Recentering-on-reversal prompts us to consider the converse—something is wrong with connectivity. Further, it prompts us to consider these very people as our window for a fuller understanding of individual's travails.

In contrast, the stakeholders tend to be relatively more aware of the system's gains. In the case of RFD, right from the beginning, rural small businesses raised the alarm against expanding the reach of urban mail-order companies. The farmers, however, were not responsive because, on the one hand, lower prices of mail-order companies benefited them, and, on the other, they begrudged rural businesses for overcharging them in

Table 9.5 Blind spots and recentering-on-reversal

	What Was the Blind Spot?	Why Did the Blind Spot Occur?	Could Recentering-on-Reversal help?	Takeaway
Postal system	Proponents of RFD sought the preservation of rural communities. Instead, it enabled information and material flows that citified them.	Preoccupation with the farmer's gains distorted the field of vision.	Decentering the farmer could have opened up a new field of vision, wherein urban interests are bigger winners.	The gains and travails of the individual and the system are not discrete but relational. Therefore, one should not only identify gains and travails with recentering-on-reversal, but also gauge their relative scales.
Education	The individual's travails were overlooked.	The nature of the social system was taken to be a given—the individual had to be tutored to fit in.	By drawing attention to the individual's travails, it could have energized the question: should children be molded to enhance the well-being of the society, or should the society change to enhance the flourishing of children?	Unarticulated travails ease the implementation of the dominant model.
Electrification	Utilities took rural electrification to be economically unviable.	Utilities saw rural networks as linear extensions of their urban networks.	The proponents decentered the urban areas and center-staged rural communities. In this changed mindscape, lateral connections among rural communities gained salience—the area coverage strategy.	Recentering-on-reversal can help develop solutions customized for the conditions of the periphery.

	Proponents focused solely on the gains of rural electrification, dreaming of decentralizing industry to rural areas and stemming depopulation.	Proponents were optimistically biased toward local solutions enabled by electrification when they needed a global view with its attendant implications—both positive and negative.	The proponents' view was limited to the end nodes—sockets that would power local solutions. Center-staging of transmission lines would have opened another view—tighter integration with the metropolitan economy.	It is prudent to undertake recentering-on-reversal on one's own perspective when adversaries do not do so.
Telephony	Bell Telephone Company focused on population density and believed that rural demand could not justify investment in rural telephony. Independents, once themselves oddities at the margins, later scoffed at an oddity at their own margins—barbwire-based lines.	Bell did not understand the intensity of demand: rural homesteads are not just places of residence, but also of production. Independents fell into the trap that centers typically fall into when viewing their margins.	Farmers undertook recentering-on-reversal with actions on the ground—they erected jerry-rigged networks that connected neighbors. Independents' dismissal of barbwire-based lines did not backfire on them. But this attitude was problematic, as we never know which oddity could be opportune for recentering-on-reversal—another entity could capitalize on it.	Recentering-on-reversal can be a product of actions on the ground alone—without a deliberate conceptual strategy. A difficult bit on the margins is not an oddity for those situated on the margins—it is their everyday reality. When empowered, these people will undertake recentering-on-reversal on their own. A position of centrality engenders a proclivity to be dismissive of the oddities on the margins. We need to guard against this and proactively counteract it.

(continued)

Table 9.5 (continued)

	What Was the Blind Spot?	Why Did the Blind Spot Occur?	Could Recentering-on-Reversal help?	Takeaway
Public libraries	Proponents of public libraries sought to bring everyone, especially the disadvantaged, into the fold. Part of the population (one-fifth presently) never uses a public library at all.	Scholars point to middle-class sensibilities, teacher tonality, and other alienating factors.	This domain is ripe for recentering-on-reversal. It has a clear generative metaphor (teacher-student relationship) and a persisting difficult bit at the margins—the part of the population that never uses a public library.	Acknowledgment of the difficult bit (step 1 in a three-step process; see figure 1.2)—the existence of the never-users—does not necessarily lead to subsequent steps. Recentering-on-reversal is not purely a seminar exercise. (Since we know little about never-users, we need to develop a deeper understanding of who they are and what their perspective is before we can engage in recentering-on-reversal.)
Broadcasting	Developers of radio—governments and corporations—did not understand broadcasting.	Developers saw radio as a wireless extension of the telegraph—that is, the wireless telegraph. In their eyes, radio waves' tendency to spread was a nuisance.	Developers' field of vision was limited by the telegraph metaphor. Amateurs were drawn to developers' difficult bit—the tendency of radio waves to spread. They started playing music over their jerry-rigged transmitters for fun.	Recentering-on-reversal can occur without a deliberate decision to do so.

	The belief that the problem of interference made the formation of a unitary federal system unavoidable.	The development of broadcasting regulation was based on the unquestioned assumption that for the formation of a broadcasting system, all interference had to be removed.	Decentering of interference as the central organizing issue opens up new possibilities. Such systems are messier and accommodate local idiosyncrasies.	Recentering-on-reversal opens up unquestioned assumptions for interrogation.
Internet	Connection is taken to be inherently good, and, conversely, disconnection is inherently bad.	The travails of connection are not considered.	Recentering on people who can afford to connect but willfully choose not to could provide keen insights into the travails of connectivity.	An understanding of the individual's travails could have prompted the development of more thought-out systems.

the past. From the standpoint of preserving their communities, both lost as the information and material flows enabled by RFD citified them. In the larger debate on RFD, while the system's gain was considered, the *relative* scale of the individual's and system's gains was not. On the other hand, in the case of rural electrification, the system's gain was not given due consideration. While the proponents dreamed of decentralization of industry to rural areas, electrification tightened the integration with the metropolitan economy, and, furthermore, facilitated depopulation by reducing the need for rural labor. These blind spots occurred because the proponents were animated by and focused on the individual's gains—mail for the farmer, electricity for the farmer. Decentering of the individual's gains would have allowed a better understanding of the *relative* scale of the system's gains and its implications for rural communities, but not ensured it.

Our cases also provide insights into the different ways in which recentering-on-reversal could occur. In the case of electrification, it began as a paper-and-pencil exercise. Socially conscious engineers, eager to find a way of creating economically self-sustaining rural networks, as opposed to highly profitable ones, looked afresh at the practices of the utilities (network planning, cost accounting, and others) and interrogated their assumptions. Based on this analysis, they developed the area coverage strategy, and thereafter, they implemented it on the ground and demonstrated its viability.

In the case of rural telephony, recentering-on-reversal was a product of actions on the ground alone, without a deliberate conceptual strategy. In the face of Bell Telephone's dogged refusal to extend service to rural areas, farmers started erecting their own jerry-rigged networks. They did not set out to execute a strategy, let alone perform recentering-on-reversal. They simply took it upon themselves to satisfy their own needs, stringing wires to connect with neighbors. In our analytical terms, a difficult bit at the margins is not an oddity for those situated on the margins—it is their everyday reality. When empowered, they could undertake recentering-on-reversal quite naturally and develop solutions customized for the conditions on the margins.

We see an interesting resonance in the case of radio. The developers of radio—governments and corporations—saw it as a wireless extension of the telegraph. In their eyes, the tendency of radio waves to spread was a nuisance. Radio amateurs, on the other hand, were drawn to this difficult bit. They started playing music over their jerry-rigged transmitters for fun, attracting audiences. Subsequently, a major corporation, Westinghouse, saw the commercial potential of broadcasting. Later, as the broadcasting industry grew, the federal government brought it under a unitary regulatory system on the basis of a totalizing logic—there should be no interference at all. As we show, the decentering of interference as the central organizing issue opens up new possibilities. For instance, if instead of zero tolerance for interference, we accept some interference below a certain threshold (e.g., only in the night, when radio waves propagate farther), we have more options for organizing the broadcasting system. Such systems are messier, but also much more accommodating of local idiosyncrasies.

The public library is an unusual and educating case. The public library movement sought to bring everyone, especially the disadvantaged, into the fold. Yet a significant part of the population (currently one-fifth) never uses a public library. Interestingly, public libraries recognize this difficult bit—in fact, they openly acknowledge it, typically in their annual reports. Yet they continue to focus almost entirely on people who do use libraries. One could debate the mission of public libraries and whether it calls on them to seek out never-users.

From our standpoint, three points are noteworthy. One, acknowledgment of the difficult bit (step 1 in a three-step process, as shown in figure 1.2 in chapter 1) does not necessarily lead to subsequent steps in the recentering-on-reversal process—libraries acknowledge the high proportion of never-users but remain focused on the users. Two, this domain, public libraries, is ripe for recentering-on-reversal. It has a clear, generative metaphor (the teacher-student relationship) and a persisting difficult bit at the margins—the population that never uses the public library. Three, since we know little about never-users in that library surveys have focused on users, we need to develop a deeper understanding of who they are and what their perspective is before we can engage in recentering-on-reversal. In effect, it cannot be purely a seminar exercise.

Finally, we also have a cautionary tale among our cases. Independent telephone companies, once oddities themselves, later scoffed at an oddity at their own margins—barbwire-based telephone lines. As subsequent developments show, the independents' dismissal of barbwire-based lines did not backfire on them. But this attitude was problematic. All oddities should be examined, as we never know which one could be a fulcrum for recentering-on-reversal, an opportunity that could also be capitalized on by another entity. Conversely, if another entity does not do so, we could benefit from performing recentering-on-reversal on our own perspective.

As we saw in the case of electrification, the proponents focused solely on the gains of rural electrification, dreaming of decentralization of industry to rural areas and stemming of depopulation. Their view was limited to the end nodes—sockets that would power local solutions. Center-staging of transmission lines would have opened another view—tighter integration with the metropolitan economy. In the end, rural electrification neither decentralized industry nor stemmed depopulation. The proponents would have been better off if they had performed a recentering-on-reversal from their own perspective.

The abovementioned examples show that the applicability and payoff of recentering-on-reversal is contingent on the context. We have cataloged the few cases that we have seen so far. With more cases in the future, we will be able to more fully understand how to employ recentering-on-reversal most effectively.

In sum, recentering-on-reversal is a very specific analytical strategy. It could be brought about in various ways. We have discussed three ways: (1) incremental improvisations on the ground, (2) deliberate analyses prompted by the press of practical problems, and (3) seminar exercises. It helps identify blind spots, prompts counterintuitive thinking, and generates new understandings. At the same time, every oddity on the margins of the established order cannot be a pivot for a reversal, as we saw in the case of barbwire-based telephone lines. But then again, trying recentering-on-reversal typically costs little. It is essentially an exercise on the conceptual plane. Moreover, when recentering-on-reversal opens new views, the payoff is big. We should, therefore, never dismiss oddities on the margins that come to our attention—instead, we should keep an eye out for them.

FINAL REFLECTION

We are now at the end of a twenty-five-year journey set off by an unease. At first, the unease concerned burgeoning research on universal access for telecommunications services. On the one hand, in the first author's view, researchers were producing copious papers on universal access, and, on the other, they were circling in narrow circles, conceptually speaking. In a nutshell, upon the spread of a new technology, researchers would identify the have nots, chart the gap, and call for universal access initiatives, and then, when the next new technology arrived, researchers would repeat the cycle. Seeking to broaden our lens, to develop a different perspective, the first author delved into the development of universal education. This yielded the first insight for our present project: universal education became a priority more than half a century after independence, only after it started suiting the interests of the powers that be, discussed at greater depth in an article entitled "Universal Service: Prosaic Motives and Great Ideals" (Sawhney 1994). Consequently, the unease deepened as universal access research continued to focus on the have nots of new technologies, when larger issues were play.

To cut a long story short, in 2003, the first author wrote a paper entitled "Universal Service Expansion: Two Perspectives," which later became the stepping-stone for the present project (Sawhney 2003). This paper argued that the benefits of universal service expansion should be looked at not only from the standpoint of the individual, but also the system. From today's vantage point, the examples provided in this paper look quaint: touch-tone phones enable banks to automate account management, e-tickets enable airlines to cut operational costs, and information services enable governments to deliver welfare benefits electronically. The paper ended by saying:

> Universal service is a far bigger phenomenon than a large-scale welfare project. It involves flows of billions of dollars. Any enterprise that touches on such large amounts of money is likely to generate complex agendas of varying shades of altruism and self-interest among the different stakeholders. Of course, the recipients of subsidized service benefit from the monies directed their way. But who else benefits beyond the generic network externalities? How does their vested interest shape what is said and done about universal service? Universal service does have a welfare component. But that is not the complete story. (Sawhney 2003, 330–331)

This line of thought, lingering on the back burner, was reenergized by Siobhan Stevenson's 2009 paper, "Digital Divide: A Discursive Move Away from the Real Inequities." In this paper, Stevenson (2009) says that the digital divide has become "an industry of global proportion" (1). Researchers study access inequities along myriad social dimensions, such as income, race, sex, age, education, geography, health, physical and cognitive abilities, and motivation. The government agencies and foundations provide monies for data collection. Policymakers and administrators discuss reams of tables, graphs, and figures so produced, and then allocate resources for digital divide programs, which engage academics, activists, and employees of governmental and nongovernmental organizations. According to Stevenson (2009), all this activity makes "social inequality and the wealth gap a technical issue and concentrated on America's most disenfranchised" (10). In effect, the problem gets reduced to increasing percentages of people of different types with access to different types of technologies.[2] Stevenson says that all this activity is a distraction. It keeps us away from a fuller understanding of the digital divide and, consequently, meaningful action. We end up fiddling with numbers-laden technocratic exercises without addressing the causes of the digital divide, which are of a structural nature.

A few years later, around 2012, a chance conversation between the coauthors led to a discussion on Stevenson's paper, a sharing of mutual enthusiasm, and eventually the present project. Stevenson shows how obscurations hamper us from developing a fuller understanding of universal access. For our part, we seek to lay bare coalitions that bring about universal access, to engender an honest conversation, and flag blind spots, to elevate the conversation to the greatest possible extent.

With regard to the latter, we catalog the blind spots identified in the course of our research, and we hope that others will build on it in the future. Beyond specific blind spots, we hope that the very creation of such a catalog will raise the salience of blind spots in general. We will now bring our discussion on blind spots to a conclusion by reflecting on a blind spot of our times that is consequential for not just system development.

Our connectivity-centric thinking lionizes the power of the link, taken to be a transformative power. In our analytical vocabulary, our connectivity-centric perspective center-stages the links, fading away the salience of the nodes. If we decenter the links and center-stage the nodes, we get a very different perspective that throws the nodes into sharp relief. We are alerted to differences in their relative power—power writ large (economic, political, cultural, etc.). Herein, the establishment of a link activates all the facets of power possessed by the nodes brought into networked proximity. Connectivity is no longer benign, let alone inherently good.[3]

We saw this dynamic play out vividly in the chapters of this book on the postal system, electrification, and telephony. As we develop universal access policies for the internet, we need to understand connectivity in all its complexities—not just gains, but also travails. For instance, shouldn't disadvantaged persons with subsidized connectivity also have their privacy issues given due consideration—from their standpoint?

With regard to the former, we offer a simple typology that brings into attention not only the individual's gains and the system's travails, but also the individual's travails and the system's gains. In our estimation, its power lies in its simplicity. We hope that it will prompt people to look afresh at universal access issues of import to them and ask profound questions that are beyond our present mindscape.

We have been celebrating the logic of inclusion, and, correspondingly, aspects of universal access that further "inclusiveness," "access," and "benefit" are clearly articulated, openly implemented, and showcased. But this is not the complete story, as universal access is a humanitarian project only to a certain degree. Universal access also has an instrumental, if not dark, side. It is also a binding project, which brings everybody into the system's fold. This logic of binding works through conscription, conversion, and travail—things that would give occasion for disquiet if we attend to them. These aspects of universal access are not articulated and are implemented quietly. They need to be made visible. We would then be pressed to also weigh in the individual's travails and the system's gains and, accordingly, develop and implement better thought-through universal access policies.

BEYOND BORDERS: CORRECTIVE TO AN AMERICAN EXPORT

Until the 1980s, access-related policy discourse the world over centered on the notion of public service. In the US, it centered on the notion of universal service, later evolving to universal access. They were born of different traditions.

At that time, countries typically had PTTs—state-owned entities providing post, telegraph, and telephone services. Japan, Canada, and the US were exceptions to varying degrees. Japan had separate entities for electronic communications (Nippon Telegraph and Telephone) and regular mail (Postal Services Agency); however, both were state-owned. Canada had a state-owned postal service and a complex mix of private and state-owned entities for telephone and related services. The US also had a state-owned postal service, but it had privately owned telephone companies. The notion of universal service developed in the world of the latter.

Broadly, both public service and universal service uphold the social obligations of service providers (e.g., reliability, quality, price, equality of treatment, and access). But they are quite different in character. Lucien Rapp (1996), protesting the European Union's adoption of universal service in spite of the longstanding European tradition of public service, wrote that "universal service imposes a 'from the bottom' approach, from the undertaking's point of view, and not a 'from the top' one, from the point of view of the state, which is the guarantor of the respect of Republican principles, notably equality and solidarity" (395). Put more simply, since the motivations of state-owned entities and for-profit companies are very different, the level of state and public vigilance necessary to ensure that they deliver on their social obligations is profoundly different. Public service considerations guide the everyday actions of PTTs; they, as state actors, are implicitly obligated to discharge their social obligations and answer to people's representatives. On the other hand, universal service is an explicit state intervention to get for-profit entities to discharge their social obligations. It is a corrective of sorts.

Within a couple of decades, by the turn of the century, the picture had flipped. François van der Mensbrugghe (2003) noted that "today, universal service has received statutory expression throughout the world: the existence of universal service obligations would be recognized in 138 countries" (1). This transformation accompanied a deeper transformation—privatization of the PTTs the world over, in a spectacularly short period of time. Countries' motivations for privatizing their PTTs were complex, involving factors such as the following: the rising demand of big business for customized solutions, which PTTs were slow to satisfy; convergence of telecommunications and computer technologies, which made demarcation between regulated and competitive services difficult; and governments' need for money and investment, which privatization obtained. What is important for us is that with privatization, countries the world over had a new need—a regulatory model for ensuring that for-profit companies met their social obligations. The only game in town, writ large, was the American regulatory model, which featured universal service, among other things.

In this flipped picture, our analysis of America's universal access experience, which was eye-opening in consequential ways, calls for also considering the global consequences of this American export—universal access. For instance, when only about 54 percent of Africans live in areas, let alone homes and compounds, served by piped-water supply (Howard and Han 2020), most African countries have established universal service funds for telecommunications—see table 1 in Arakpogun, Wanjiru, and Whalley (2017), which records universal service funds in thirty-four of fifty-four countries in Africa. Further, influential bodies such as the World Bank, World Trade Organization, and International Telecommunications Union actively promote universal service with funding, consultants, and training workshops. Moreover, they often join forces. For instance, the World Bank, European Commission, United Nations, and REGULATEL[4] jointly sponsored the study *New Models for Universal Access to Telecommunications Services in Latin America* (Stern and Townsend 2006). In a similar vein, on a more granular, prototype level, Michael Trucano, World Bank's global lead for innovation in education, advocating for the establishment of universal service funds, says that "the most famous, and indeed the prototypical, example of how a 'universal service fund' has been used to connect schools to the internet is the e-rate program in the United States" (2015).

We do not explore the issue of whether the global spread of universal access is good. We are moving at a more basic level, advancing the argument that an American export should be coupled with its corrective. Countries create and run their systems in distinct ways, in accordance with their particular sociohistorical context and political ethos. Consequently, on the basis of our study of universal access in the US, we cannot speak on issues and implementations specific to any other country, let alone the world at large. Generally, it is quite possible that researchers, policy analysts, and other stakeholders in other countries would find our gains and travails framework useful—it is simple, general, and abstract. But then, actual effort by such stakeholders to put it to use would be the real test. On this score, we are not even invested in the gains and travails framework per se. If it serves as a prompt to broaden the discourse with some other means our counterparts in other countries find more useful, it will have served its purpose.

NOTES

1   INTRODUCTION

1. We decided against the commonly used terms "cost" and "benefit" because of their limiting economic undertones. Thereon, we looked for ones that would help us think afresh in two ways: by being free of thought-limiting baggage and by prompting a wide-angle view. We eventually settled on "gains" and "travails."

As our use of "gain" is relatively straightforward, we will be spare and not belabor it. On the other hand, our use of "travail" calls for some discussion. Etymologically, "travail" comes to us from Latin via Old French. Its Latin root is "trepalium," an instrument of torture—"tres" (three / tri) and "palus" (stake). Since then, it has acquired milder meanings ("travel" is also its descendant). In today's French "travail" means work. In contemporary English, "travail" has a wider range of meanings: "physical or mental work, esp. of a painful or laborious nature; great effort or exertion; hardship, suffering; (now frequently in weakened use) trouble, difficulty" (Oxford English Dictionary Online, July 1, 2021). All in all, "travail" nicely satisfies both our considerations. It is an uncommon but familiar word and has almost no history of usage in policy research, at least as a term of art or anything even remotely approaching it. Further, in contemporary English, its range of meanings, all centered on some degree of suffering, help us maintain a wide angle on what the individual bears in return for universal access.

In other contexts, authors have talked of "travails of the American labor movement" (Cochran 1995, 466); "travails of a widow" (Onuegbu 2006, i); "travails of a progressive conservative" (Lurie 2012, iii); "travail of native Americans" (Weeks and Gidney 1981, iii); and "human travail of Russia" (Dawson 1945, iiii). On our part, we talk of the travail of the individual. Among the domains we focus on, we have a rare essay by columnist Henry Mencken's entitled "Travail" wherein he views the education system from the standpoint of the student. He starts his essay thus: "It always makes me melancholy to see the boys going to school. During the half hour before 9 o'clock they stagger through the square in front of my house in Baltimore with the

despondent air of New Yorkers coming up from the ferries to work . . . In the after-
noon, coming home, they leap and spring like gazelles" (Mencken 1955, 182). In the
rest of the essay, he reflects on why this is so, laying out various reasons: nature of tasks,
rules, and teachers; lack of human-level respect and connection, genuine learning, and
excitement. Our approach is similar to Mencken's in one important way and also different in an important way. We also endeavor to be comprehensive, laying out myriad
aspects of travail. However, unlike Mencken, we do not pass a judgment. We only seek to
fully understand. Further, Mencken does not delve into gains from education—it was
not part of his project. We, on the other hand, delve also into the gains from the education and other systems, in the same comprehensive manner as the travails. We use
"gains" and "travails" for their utilities explained above. However, we are not wedded
to these particular terms. If the reader prefers to use substitute terms, in our estimation, it would not detract from our analysis. One possibility would be the approach
in Jayakar (2017), where "costs" and "non-monetized costs" (e.g., psychological stress)
are considered separately, as are "benefits" and "non-monetized benefits."

2. Here, we are reminded of William James's observation, made famous by Mary
Douglas: what we consider to be "dirt" is essentially "matter out of place"—same par-
ticles on the ground outside the home would not be dirt (quoted in Douglas 1966, 165).
She added: "Dirt offends against order. Eliminating it is not a negative movement, but
a positive effort to organize the environment" (Douglas 1966, 2). In contrast, from the
perspective of reversal, dirt presents an opportunity for a positive moment in a radically different sense—an invitation to interrogate the present order and discover other
possible orders by center-staging it.

3. For the selection of such cases, Thomas identifies three routes: the local knowledge case (familiar to the researcher), the key case, and the outlier case.

4. Conceptually, "a node is a point in a network where two or more links intersect. A
link is a path between these nodes" (Carlton and Klamer 1983, 450). Physically, a node
could be a building (e.g., a post office or school) or a switching mechanism of some
sort (e.g., a telephone switch or power substation). The connections between nodes
could be provided by other systems (e.g., postal system–enabled interlibrary loans of
books), or they could be integral to the system (e.g., long-distance telephone lines).

5. Ritchie, Lewis, and Elam (2003, 88) also have a third question for human subject–
based studies: "Will the sample frame provide a sufficient number of potential participants to allow for high quality selection, particularly given that not all will be
eligible or willing to participate in the study?"

6. REA also oversaw the expansion of rural telephony, often employing strategies
that it had employed for rural electrification. Also, power-line telecommunications
technologies use electricity distribution networks to transfer information.

7. More broadly, welfare initiatives are rarely entirely humanitarian endeavors (see
Esping-Andersen 1998), something we overlook when it comes to universal access
programs.

8. The degree by which the individual is beholden to the system varies, however,
from case to case. In the case of telephony, the level of conscription is pretty high. Just
consider how the ring of the telephone often trumps a conversation with copresent

others. Moreover, what seems like an instinctual picking-up of the telephone and starting a conversation is in reality a highly conditioned behavior, a product of much tutoring. Subscribers were taught to "place the telephone fairly against the ear," "speak directly into the mouthpiece keeping the mustache out of the opening," and follow the protocol wherein the caller signals end of conversation unless a man was calling one of the "second sex" (Fischer 1992, 70–71). On the other hand, in the case of public libraries, the level of conscription is very low. It is available as a community resource, but its actual use is pretty much up to the individual.

2 POSTAL SYSTEM

1. Behringer (2006) considers the Taxis service to be the first modern postal system on two counts. One, unlike earlier postal systems restricted to state-approved users, it was open to the public upon payment of a fee. Two, it operated on a regular schedule. Moreover, the contract between the Taxis and the Hapsburgs "laid down the basic terminological framework of the modern European postal system" (341).

2. William Blackstone, the author of the influential *Commentaries on the Law of England* (1765), characterized the postal service as "most eligible method . . . of raising money" (quoted in John 1995, 26).

3. In 1852, a Michigan congressman noted that over his twenty-five years of service, not one application for a new mail route had been turned down (John 1995).

4. US Postal Service, Free Delivery, http://about.usps.com/publications/pub100 /pub100_018.htm (accessed March 25, 2016).

5. Some paid a private company to pick up their mail from the post office.

6. US Postal Service, Rural Free Delivery (RFD), http://about.usps.com/publications /pub100/pub100_020.htm (accessed March 25, 2016).

7. Some in rural America even felt that that the postage they paid was subsidizing the free postal delivery in the cities (Law 1996).

8. US Postal Service, Rural Free Delivery (RFD), http://about.usps.com/publications /pub100/pub100_020.htm (accessed March 25, 2016).

9. The skeptics included people in the post office, who had long opposed the expansion of postal service in rural areas on the grounds that it could never be self-supporting (Roper 1917).

10. Bissell continued to feel this way a decade later. In his estimation, the government had made a "great mistake" by introducing RFD. He anticipated the actual cost to far exceed the estimated one. Furthermore, he believed that it would increase the isolation of rural residents by reducing their in-person interaction with people in the nearby town—post office employees, country store owners, and their employees, and others (*New York Times* 1902b).

11. Gary was partly quoting a report of the Committee on Post Offices and Post Roads of the Fifty-Fourth Congress.

12. Patronage of all the residents of a service area was critical for the success of RFD. At times, rich residents' reluctance was a problem. Since they owned automobiles, they did not see mail pickup at the post office as a burden (*New York Times* 1908a).

13. According to Heath, the value of farmland touched by RFD increased from $2 per acre to $3 per acre (*Grand Rapids Herald* 1899).

14. They also carried printed cards with more detailed information (*New York Times* 1901d). This lasted only a short while because around 1902, newspaper and telephone became the main channels for the dissemination of weather information (Law 1996).

15. The abovementioned gains of rural people were also gains of the system. For instance, farmers' time savings with RFD also increased the productivity of the overall economy. According to one estimate, the elimination of farmers' trips to the post office was worth millions of dollars in time saved (*New York Times* 1899). Similarly, the increase in the value of land increased tax collection, wider dissemination of prices increased the efficiency of the markets, and so on.

16. Beyond its wide-ranging metaphoric import, shedding light on myriad mechanisms of modern empire building, Carey's aphorism also had a very literal rendition. The British commonly named places after imperial surveyors. Most famously, the world's highest mountain, known as Sagarmāthā in Nepali, was named Mount Everest after George Everest, surveyor general of India.

17. Often, the local postmaster would also double up as a storeowner, or conversely, a storeowner would also serve as a mail pickup station.

18. The post office facilitated this by setting up special rates for deliveries weighing less than four pounds (helpful for provisioning items such as food, coffee, tobacco, and drugs). As Postmaster Robert Wynne explained, "the system by which these packages are to be delivered is already established, and such delivery would entail no additional expense upon the department" (*New York Times* 1904, 6). A similar logic was at play, when, in 1908, P. V. De Graw, an assistant postmaster general, recommended, in his annual report, an increase in the allowable parcel weight to eleven pounds, arguing: "It can be given with the facilities now employed and would materially increase the revenues of the department" (*New York Times* 1908b, 8).

19. In 1889, after a year of limited RFD service, Sears reported selling "four suits and a watch every minute, a buggy every ten minutes and a revolver every two minutes" (Law 1996, 102).

20. Before the start of parcel post, the post office could carry packages weighing only up to four pounds. For bigger packages, rural resident's choice was either the more expensive, quicker express companies or the much cheaper, slower railroads. But the railroads charged to carry a minimum of 100 pounds. To get the best out of these options, the mail-order companies encouraged customers to come together and aggregate their orders into one big one. Sears, in its catalog, told its customers: "So that in case you only have a small order and want it to come by freight, you could have some friend buy goods at the same time, send his orders with yours, and have both orders shipped in your name" (quoted in Kielbowicz 1994b, 97). In this pre–parcel post world, in 1911, Montgomery Ward shipped 82 percent of its orders via railroads and only 8 percent via mail (Kielbowicz 1994b).

21. The notion that connectivity is inherently good was conjoined with the notion that the circulation of information is inherently good. The Congregationalist minister Leonard Bacon in 1843 asked a profound question: Why was the government subsidizing the

circulation of newspapers and not Methodist circuit riders? (John 1995). Fundamentally, he was questioning the type of information that RFD would put into circulation. He feared that it would secularize society. The fact that RFD served as a conduit for the spread of consumer culture should give us reason for pause, whether we are for or against it. The deeper lesson for us is that we need to go beyond our preoccupation with the circulation of information and pay attention to the type of information in play.

22. The post office was also okay with the carriage of other periodicals, as they carried much less advertising than did magazines.

23. In 1928, RFD constituted one-seventh of the overall cost of the postal service but carried only one-sixtieth of the overall mail (*New York Times* 1924a).

24. They are expensive to create and maintain (i.e., their fixed cost is very high). But the marginal cost associated with each use is very low, even negligible for electronic systems. In the case of such systems, the allocation of fixed costs is the determinative factor. However, it is a matter of contention because there is no determinative rationale for cost allocation itself—there are myriad rationales. Furthermore, such systems are characterized by network externalities, wherein each node increases the overall value of the system, but in ways that elude quantification. In such a setting, myriad interpretations arise—each meritorious in its own ways. For instance, Postmaster General Frank Hitchcock's statement, in 1910, reported that RFD made a loss of $28 million in the last fiscal year. Representative Sims of Tennessee took issue with this statement on two counts. One, the postmaster had not taken into account savings of $5 million from the discontinuance of old services replaced by RFD, and also the closure of post offices made redundant by it. Two, the postmaster "credited the rural service only with the collection of mail made on the routes and not with any part of the mail delivered" (*New York Times* 1910a, 2). The latter point is particularly noteworthy for our analytical exercise, as Representative Sims was criticizing the post office for overlooking the gain to the system—namely, the delivery of newspapers, magazines, mail-order packages, and other items to rural homes.

25. While this dispute produced travails for the system, for our analytic purposes, it is important to note that the argument was about who was making unfair profits. The fact that profits were being debated at all is more important than who was making them. Either way, though, they constituted gains for the system.

26. A *New York Times* article on the negative impacts of RFD on rural businesses complained: "The individual chiefly concerned is the farmer, and he has benefited" (*New York Times* 1902e, 5).

27. In fact, James Cowles, founder of the Postal Progress League, explicitly said of parcel post: "It will tend to make the whole United States one great city, within which our various transportation agencies will move to and fro like a weaver's shuttle, weaving a web of ever increasing concord and prosperity" (2012, 237).

3   EDUCATION

1. Educational institutions are also sorting mechanisms that reproduce and reinforce social stratification (Arboleda 1981; Birenbaum 1971; Lavin, Alba, and Silberstein 1981; Ornstein 1974; Rossman et al. 1975; Schrag 1971; Willingham 1970).

2. The 1563 law transferred to the state functions that earlier had been performed by guilds (e.g., wage-setting, training). The 1601 law provided monies for basic support of the poor and required them to be trained in a trade (Kotin and Aikman 1980).

3. Rhode Island's religious heterogeneity deterred finding a consensus on education (Kotin and Aikman 1980).

4. Our understanding of what constitutes adequate education changes over time, often expressed in terms of "the provision of educational opportunity which is needed in the contemporary setting to equip a child for his role as a citizen and a competitor in the labor market" (Wise 1983, 301).

5. The 1648 amendment specified that students be taught to read perfectly (Kotin and Aikman 1980).

6. The New Haven Colony, established in 1638, was merged with the Connecticut Colony in 1665.

7. In this period, four major factors undermined universal education. One, increasing conflict between the colonial legislatures and British imperial authorities often led to the repealing of laws. Two, economic hardship was created by the Indian wars. Three, increasing geographic spread of the population made enforcement difficult, and at the same time, the need for labor in frontier settlements increased the opportunity cost of education. Four, there was a decline in motivation to teach Scripture to children, corresponding with the decline in the importance of religion (Kotin and Aikman 1980).

8. In effect, universal suffrage would be meaningless without universal education (Adams 1875; Boone 1889; Carlton 1906).

9. The founders, concerned with the fragilities of the newly born country, saw universal education as a vehicle for perfecting the union (Messerli 1967). For instance, in 1786 Benjamin Rush wrote of "producing one general and uniform system of education" that would "render the mass of the people more homogeneous and thereby fit them more easily for uniform and peaceable government" (Rush 1965, 10). George Washington carried this theme into his will, where he expressed a desire "to see a plan devised on a liberal scale which would have a tendency to spread systematic ideas through all parts of the rising Empire, thereby to do away local attachments and state prejudices" (Thomas 1967, 3).

10. All education systems are marked by a tension between "the good of society" and "the good of the individual" (Rudolph 1965, xv). The founders were biased toward the former.

11. More than a century later, Chief Justice William Brennan even talked about "the economic function (of the education system in) keeping children of certain ages off the labor market and in school" (*Wisconsin v. Yoder* 1972).

12. It echoed Benjamin Rush's argument in 1786 that education was necessary for making citizens "republican machines." In 1849, supporters of New York's 1849 School Law also echoed this point: "We hold, therefore, that our present school tax is not imposed on the rich for the benefit of the poor; but imposed on the whole State for the benefit of the State" (Carlton 1906, 48–49).

13. Thaddeus Stevens, U.S. representative from Pennsylvania, pointedly noted that people paid taxes for jails even when they did not use them (Stevens 1964).

14. In effect, people of this belief reject the notion that education is a public good. In a striking contrast, advocating for universal education, Strong (1963, 99) said: "Why does the state take money from your pocket to educate my child? Not on the ground that an education is a good thing for him, but on the ground that his ignorance would be dangerous for the state." In this formulation, the benefit that accrues to the individual is rendered only incidental. The rich pay taxes for universal education not for the benefit of the disadvantaged children but for themselves, as their education is important for the society as a whole, bringing about productivity gains and the reduction of problems such as poverty, intemperance, and social unrest (Carlton 1906).

15. Has centralization increased? For many conservatives, the establishment of the US Department of Education in 1979 epitomizes centralization. But then, the US continues to have more than 3,000 school districts that jealously guard their local autonomy. The reality is perhaps quite complex, as increased centralization does not necessarily mean that the system is centralized.

16. Benjamin Peers, the author of a mid-nineteenth-century tract advocating universal education, warned: "In a society where everyman may do pretty much as he pleases, it is of the utmost importance that its members be so educated that they shall choose to do right" (Jackson 1941, 66).

17. Jackson (1941) summed up the importance of universal education from the standpoint of the body politic as follows: "The American experiment, basically desirable and an improvement over anything previously known, yet bore seeds of destruction whose shoots could be held within bounds only by education. Of them, universal suffrage was the most obvious" (65).

18. In the 1960s and 1970s, homeschooling was against the law and largely confined to hippies and back-to-the-landers, involving about 10,000 to 15,000 children. Only later, in 1993, did it become legal in all fifty states (Heidenry 2011). Today, the practitioners of homeschooling are largely religious-minded parents, who like the hippies and back-to-the-landers want to keep their children out of the system-administered public schools. The latest available statistics, for 2016, show that 1.67 million children are presently being homeschooled (National Center for Education Statistics 2018).

19. On the production of industrial workers, the French engineer-sociologist Fédéric Le Play memorably said, "The most important product of The Mines is The Miner" (Schrage 2000, 13).

20. Wayland is basically talking about "network externality," a term typically not used in education. It, however, is noteworthy for us, as our project includes many systems wherein network externality is a major issue.

21. On the administration of aptitude tests, the military found that twice as many persons had the capacity to attend college than those actually enrolled (Bowles 1966; Munday and Rever 1971; Willingham 1970). Furthermore, after the war, studies

indicated that the increased taxes paid by the beneficiaries of the G.I. Bill exceeded the cost of the program. In effect, the taxpayer incurred no net cost for the education provided to veterans via the G.I. Bill (Educational Policies Commission 1964). These war-related experiences set the stage for the subsequent discussion.

22. The anti-elitist atmosphere of the 1960s and 1970s energized activists to push for universal access to higher education. However, in marked contrast to the earlier expansion of universal education, a knotty question arose: Should the beneficiary of taxpayer largess have the talent required to make good use of it? Scholars, who see the education system as a sorting mechanism that perpetuates social inequities, argue that the focus should not be on picking winners, but on cultivating the maximum potential of every student who enters higher education (Karabel 1972; Rossman et al. 1975). With such a philosophy, the City University of New York offered admission to all students who ranked in the top half of their graduating class and also directed its efforts toward remedial programs for weaker students (Lavin et al. 1981). On the other hand, California created a three-tiered higher education program, with the University of California system for the top eighth of high school graduates, the California State University system for the top third, and community colleges for all high school graduates. Within this tiered system, it offered opportunities for upward mobility (Jaffe and Adams 1972).

23. While there has been much talk of providing tuition-free access to two years of college since the publication in 1947 of the President's Commission on Higher Education report *Higher Education for American Democracy*, little headway has been made over the last several decades.

24. It offered a last-dollar scholarship, wherein students applied for federal student aid (e.g., Pell Grants) and Tennessee covered the shortfall (Kelderman 2014; Pérez-Peña 2014; Semuels 2015).

25. This line of thinking persists. Half a century later, in 1896, Woodrow Wilson, when president of Princeton University, said at the university's sesquicentennial: "The object of education is not merely to draw out the powers of the individual mind: It is rather its right object to draw all minds to a proper adjustment to the physical and social world in which they are to have their life and their development: to enlighten, strengthen and make fit" (Kolowich 2016, A28).

4 ELECTRIFICATION

1. Describing the dramatic contrast between electrified towns and the nonelectrified rural areas, Marvin (1988, 164), writes: "When a small Illinois town installed electric lights, outlying farmers observing the glow were convinced that the town was on fire, and raced in their wagons to help."

2. Edison was the first to conceive of the electric utility model, seeing the future of the sale of electricity generated from a central power plant. J. P. Morgan, the largest stockholder of Edison Electric Illuminating Company, wanted Edison to focus on the manufacture and sale of electric appliances. With regard to central power plants, Morgan thought that they were too expensive and were not good business

from a profitability standpoint. Eventually, Edison invested his own money and pressed forward with central power plants (Rudolph and Ridley 1986). Interestingly, in the case of radio as well, Westinghouse and other companies saw business opportunities mainly in the sale of radio equipment at first, only later realizing that broadcasting by itself could be a business (Davis 1930; Douglas 1987b).

3. Marvin (1988, 18) thus retells the story of Edison soliciting a millionaire for the provision of electric lights for his house: "During the conversation, the millionaire asked whether Edison could install an electric motor to run the steam engine that operated his passenger elevator."

4. According to Hughes's conceptualization, the network development process has five phases: (1) the invention and development of a system (with inventor-entrepreneurs playing a dominant role); (2) the transfer of technology from one region and society to another; (3) the scaling-up of the network and the emergence and resolution of critical problems or reverse salients (with researchers and engineers playing a dominant role); (4) the system acquiring momentum (with managers playing a central role); and (5) qualitative change taking place in critical problems in technical and organizational areas (with the rise of financiers and consulting engineers).

5. Even here, they were concerned that few farming activities required continuous supply of electricity over the entire year—that is, across seasons (Meinig 2004).

6. Sad irons (metal plates heated on stoves) were precursors of electric irons.

7. *Oxford Dictionaries* (Lexico, https://www.lexico.com/en/definition/sad, accessed August 1, 2019). We would like to thank an anonymous reviewer for an edifying comment on the etymology of "sad."

8. He became the first administrator of the REA.

9. Morris Cooke never bought into the math employed by the utilities to estimate the cost of providing electricity to rural areas. The variance in the calculations of different utilities deepened his suspicion. For instance, an Illinois Commerce Commission study found that estimates of different utilities for the material costs of rural electrification ranged, for no clear reason, from $365 to $1,000 per mile. Similarly, the costs of labor, capital, insurance, overhead, and transportation varied, with final costs ranging from $500 to $2,000 per mile (*Electrical World* 1931). Such studies suggested that "there can be no strict allocation of costs. It is all a matter of judgment and experts differ" (Burritt 1931, 697). Moreover, he felt that technological advances had changed the economics of electricity distribution. In the 1890s, electricity was basically used for lighting in the evening. Here, the utilities had to recover their costs over a short period of time, resulting in high rates. Since then, applications of electricity proliferated, flattening the demand curve over the day. Cooke argued that an increase in quantum of demand and its flattened pattern had changed the cost structure. With these changed parameters, he argued, we need to rethink the fundamental assumptions and the methods for estimating costs (Childs 1974). Essentially, Cooke put his finger on the slippery assumptions that plague cost allocations in networked infrastructures, wherein high fixed costs need to be recovered somehow, and there are many different ways of doing so.

10. Later, REA also engaged in the development of rural telephony. In 1949, Congress amended the Rural Electrification Act to authorize REA to advance loans for the development of rural telephony (Rural Electrification Administration 1990).

11. At first, REA provided loans at a rate that covered the government's cost, fluctuating with the market. In 1944, REA fixed the rate at 2 percent, the cost incurred by the government. Later, when the interest rate rose in capital markets, the government started subsidizing REA loans. REA maintained the 2 percent rate until 1972, even though the government's cost had hovered at over 5 percent for several years (reaching a high of 5.986 percent in 1970). Even in 1973, when interest rates started fluctuating again, REA borrowers were charged subsidized interest rates—3.72 percent, with the cost to government being 6.129 percent. Such subsidies continued over the years. For instance, a decade later, in 1983, the rate was 4.90 percent and the cost to the government was 10.850 percent (Rural Electrification Administration 1990).

12. Among other things, REA offered promotional rate schedules, wherein per-unit costs decreased with increased consumption (Cooke 1935).

13. In the case of refrigerators, such efforts brought down costs by 20–25 percent (Christie 1983).

14. When the area coverage approach started showing success, the utilities sought to stymie its spread to new areas by building so-called spite lines. Their intent and calculation were as follows:

The pattern was fairly uniform. Here is a small cooperative in a moderately prosperous rural community. The plans for the project submitted to Washington provide for about 26 miles of lines to serve only about 100 members. The local power company hears that the line is about to be approved. Promptly the company drives a seven-mile power line through the middle of the project. In this way it picks up 35 of the cooperative's best prospective customers. Without these 35 customers the project is not practicable. Consequently REA is forced to refuse the loan. Then the 65 other farm households come to the company with the request to be hooked up, too. And the company replies that it is not profitable to serve any others than those already on the compact seven-mile line. Therefore, 65 households that might have had electricity are forced to go without it (Childs 1974, 50).

After the construction of a spite line, if the community still made an effort to create a cooperative, the utility tied it up in the courts, as REA was prohibited by law from engaging in areas already served by a private company (Childs 1974).

15. Initially, many people in rural areas were also bewildered by electricity. For instance, in Iowa, a local newspaper wondered whether cows would know when to sleep after the introduction of electric light, writing: "Is there a town over the broad earth where cows run loose under electric lights?" (quoted in Marvin 1988, 119).

16. Interestingly, rural mail service was also expected to stanch the depopulation of rural areas (Roper 1917). See our discussion of this in chapter 2.

17. Electricity was also expected to make small, noncash-crop farms viable, diversifying agriculture (Deutsch 1944).

18. Some even felt that at least one male child should stay on the farm (*Rural Electrification News* 1936a).

19. See steps 1 and 2 in figure 1.2.

20. See step 3 in figure 1.2.

21. There were many signs of latent demand in rural areas. Farmers were using their automobiles as sources of stationary power for turning farm machinery, powering sawmills, pumping water, and myriad other applications. Its wide-ranging use inspired a poem with the opening line: "The auto on the farm arose / Before dawn at four. / It milked the cows and washed the clothes / And finished every chore" (quoted in Kline 2000, 74). Some farmers even tapped into electric power lines for interurban trolleys if they passed through the area near their farm (Kline 2000).

22. Handwritten note by John Carmody, John Carmody's papers, Franklin D. Roosevelt Presidential Library, available at http://newdeal.feri.org/tva/tva13.htm (accessed May 1, 2017).

23. Electricity did enhance the rural economy in a number of areas, including canneries, cotton gins, flour and other mills, grain elevators, sawmills, stockyards, and lumber mills (Slattery 1940). But the impact was not as grand as imagined and certainly was inadequate to compensate for other factors that contributed to the depopulation of rural areas.

24. Similarly, Lilienthal (1939) argued, "Electricity is a flexible and mobile force. It can move out, coursing over transmission networks, to seek the worker at the forest, the quarry, and the mine" (58–59).

25. The difference between the nodal and relational perspectives comes out particularly strongly in the contrasting neologisms coined in the US and France in the late 1970s for the then-emerging phenomenon of networked computing—compunications and telematics, respectively. The American neologism focused on the computer, the node, whereas the French neologism focused on telecommunications, the connections between nodes (Bell 1980). This difference is significant. The strong suit of the US is computers, which is foregrounded in compunications. The French, seeking to stem American dominance, figured that they faced a far more uphill battle in the realm of computers than that of telecommunications, where governments play a critical role and can use "their trump card, which is to decree" (Nora and Minc 1980, 7). They strategically chose to focus on telecommunications instead.

5   TELEPHONY

1. Interestingly, in this emerging landscape, Claude Shannon, the originator of information theory, as a child in a small community in Michigan in the 1920s, jerry-rigged a barbwire-based line to his friend a half-mile away, which they used to send each other telegraph messages in Morse code. We would like to thank an anonymous reviewer for sharing with us this bit of history and directing us to Gleick (2012).

2. Bell maintained the telephones, and the purchaser bought and installed the wires. The charge for each additional phone was $10 per year (Brock 1981, Brooks 1976).

3. Western Union expanded faster—it had much deeper pockets. It also secured competitive advantage by threatening to deny telegraph service to establishments

(e.g., newspapers and hotels) that procured telephone service from its rival (Brock 1981).

4. In 1883, Bell asked its affiliates in a survey: "Is it desirable and what would be the most practical way, to provide a service which would be in the reach of families, etc.?" (Fischer 1992, 41). We see here that Bell was not sure whether residential service was "desirable" in urban areas, and service in rural areas was not even a consideration.

5. The first key Bell patent expired in 1893 and the second one in 1894 (Brock 1981).

6. In Indianapolis and Toledo, Ohio, the independents charged as low as half Bell's rate (Brooks 1976).

7. The independents did try to cooperate and build their own long-distance network, but their efforts petered out.

8. Independents' integration with the Bell-centric system was tightened when the government took over both the telephone and telegraph systems in 1918 as a war measure. President Woodrow Wilson's proclamation decreed control of "each and every telegraph and telephone system, and every part thereof, within the jurisdiction of the United States, including all equipment thereof and appurtenances thereto" (Brook 1976, 151). The post office appointed a unified Wire Control Board, which in turn appointed and oversaw a telephone and telegraph operating board. Among other things, the government benefited the industry by instituting price increases, which the industry was unable to do on its own. Moreover, "the wartime experience of coordination between AT&T and the independents accelerated the unification of the industry" (Fischer 1992, 50). The control reverted to the private companies in 1919, about nine months after the end of World War 1 (Brooks 1976).

9. The US Census Bureau started collecting data on households with telephone service in 1920 (Bureau of the Census 1976).

10. To expand coverage further in rural areas, in 1949, the US Congress amended the Rural Electrification Act of 1936 and authorized the Rural Electrification Administration (REA) to issue loans and provide advisory services for rural telephony.

11. The data on the users was as follows: wives (60 percent), husbands (23 percent), and others (17 percent).

12. In 1901, the *Washington Press*, published in Iowa, wrote:

We are glad to see 'phones threading country air as thick as gossamer in October skies . . . for these are wholesome, healthy nerves keeping the country population in vital touch with the rest of the world. Social force is good tonic (Atwood 1984, 82).

This was clearly an overstatement. While less poetic than "gossamer in October skies," the proponent's instrumental expectation that the telephone would help resolve the "rural problem" was no less sublime (Brooks 1976; Carey 1989). The reality is that connectivity is a double-edged sword, something we, enthralled with technology, tend to forget. This is very evident in the story of rural telephony.

13. See steps 1 and 2 in figure 1.2.

14. See step 3 in figure 1.2.

15. If there was a unifying point, it was the independents' opposition to the "octopus Bell." On this front, they underestimated the complexities of scaling up their own networks, the competitive advantage that Bell acquired with its long-distance network, and the difficulties of coming together and creating an alternative to Bell's long-distance network.

6 PUBLIC LIBRARIES

1. The Scoville Memorial Library, https://connecticuthistory.org/the-scoville-memorial-library/ (accessed October 11, 2018)

2. Peterborough Town Library, http://peterboroughtownlibrary.org/history-and-renovation-9330/location/peterborough (accessed October 11, 2018).

3. Wayland Free Public Library, https://waylandlibrary.org/aboutus/about-wpl/history/ (accessed October 11, 2018).

4. Wayland was the benefactor of the abovementioned $500 gift for the establishment of the public library in Wayland, Massachusetts.

5. In 1876, this threshold of 300 volumes was important for differentiating substantive libraries from those that were libraries in name only. For instance, in the midnineteenth century, Indiana and Michigan mandated that communities establish township libraries. Consequently, during the 1840s and 1850s, communities established nearly 1,600 township libraries in the two states—but nearly all of them fizzled out; only 81 attained the stature of a public library (Kevane and Sundstrom 2014).

6. American Library Association, Number of Libraries in the United States, https://libguides.ala.org/numberoflibraries (accessed June 13, 2019).

7. For his part, Carnegie, a rigidly moralistic individual inspired by writings on social Darwinism, sought "to give the 'best and most aspiring poor' the opportunity to improve; the not so good and less aspiring be damned" (Harris 1975, 15).

8. It was also in this spirit of Americanization, many years later, that the American Library Association created the Committee on Work with the Foreign Born.

9. The state averages range from 0.6 miles in the District of Columbia (technically a federal district) to 7.3 miles in Arkansas, which is an outlier—the figure for Montana, the state with the second-biggest average distance, is 4.2 miles (Donnelly 2015).

10. Schmitz, Rogers, Phillips, and Paschal (1995) report how discussions, which included participation from the homeless, on Santa Monica's Public Electronic Network (PEN) gave rise to the SHWASHLOCK (Showers, WASHers & LOCKers) project for the provision of showers, washers, lockers, and access to a laundromat for the homeless.

11. The public library's engagement with social work, a responsibility of other institutions, is controversial. For our part, we are not praising or criticizing libraries for this work. Our approach is agnostic and forensic. We just chart all the gains and travails to the individual and the system stemming from the public library, the way that they exist and function in the world.

12. See the unnumbered table on page 15 of the Pew report for the corresponding percentages for females, non-Hispanic whites, parental status, geographical location, and various age, income, and education groups.

13. The library, unlike education, does not come under the purview of truancy laws requiring compulsory attendance. This is one reason why the libraries have not elicited deep criticism, such as Michel Foucault's critiques of schools.

14. See Knuth (2003) for another library-related project that moves the focus from the individual to the system. In this project, Knuth seeks a fuller account of why we periodically see large-scale destruction of books and libraries. After describing "book burning" during Britain's Protestant Reformation and the French Revolution, she observes: "Bibliophiles quail at such daunting figures, while those who believe that book destruction can be positive point out that the reform activities weren't entirely antagonistic to the cause of libraries. As a result of forced dissolution, private and religious collections often became the property of the state and, ultimately, more accessible to the general public" (Knuth 2003, 24). For example, France used the eight million books confiscated by the revolutionaries to form a network of municipal libraries.

15. This survey was of residents of communities with populations of 300,000 or fewer.

16. Critics tend to undervalue the educational, communal, and uplifting role that public libraries can play for people from different walks of life. This might take different forms and shapes, but it is hard to dispute its *potential* for engagement and empowerment.

17. See steps 1 and 2, figure 1.2.

18. See step 3, figure 1.2.

19. See McNeely and Wolverton (2008) and Vaidhyanathan (2009) for other approaches for reenvisioning the public library.

7   BROADCASTING

1. We take the term "un-order" from Kurtz and Snowden (2003, 465), who explain it as follows:

Un-order is not the lack of order, but a different kind of order, one not often considered but just as legitimate in its own way. Here we deliberately use the prefix "un-" not in its standard sense as "opposite of" but in the less common sense of conveying a paradox, connoting two things that are different but in another sense the same. Bram Stoker used this meaning to great effect in 1897 with the word "undead," which means neither dead nor alive but something similar to both and different from both.

2. When the history of broadcasting is evoked, it is used to argue against the need for policy intervention to further universal access. The argument here is that occurrence of gaps is a natural outcome of differences in wealth, income, age, education, and other such factors. Moreover, such gaps are transitory, since they will disappear as the prices decline with mass production and technological advances. In effect, there is no need for policy intervention. As also noted in chapter 8, on the internet, this argument was memorably cast when Michael Powell, then-chairman of the

Federal Communications Commission (FCC), compared the "digital divide" to a "Mercedes divide" (*Washington Post* 2001).

In this view, universal access policies are relics of the telephony history. According to Compaine (2001b), "notions of the federal government being responsible for providing digital access to all Americans is therefore derived as an extension of the 'telephone gap' of the 1930s" (102). In a world of an ever-accelerating pace of technological change, it is no longer relevant. In fact, Browning (1996) issued a call to "bury universal service—to bury it slowly, gently, and with great care to preserve both its spirit and its many achievements."

3. KDKA was not the first broadcasting station per se. Amateurs were already commonly broadcasting messages, announcements, and music, among other things (Douglas 1986, 1987b). If the criterion for qualifying as a "station" is scheduled broadcasts, there were numerous such stations before KDKA came on the air (Davis 1930; Douglas 1987a; Hijiya 1992). The importance of KDKA lies in that it was the first broadcasting station established by a major corporation, Westinghouse. The story of its birth starts with a senior Westinghouse executive, Harry Davis, noticing that a company employee, Frank Conrad, an amateur radio broadcaster of "concerts," was featured in a department store's newspaper advertisements for radio sets (Barnouw 1966; Davis 1930; Sterling and Kittross 1990. This sparked Davis to see the business potential of broadcasting, and the rest is history.

4. The region names here refer to those of the US Census.

5. For instance, in 1928, while New York City, Chicago, and Los Angeles had twenty-eight, thirty-six, and twenty-four radio stations, respectively, Atlanta had only three and New Orleans only seven (Craig 2004a). Rural America, where 44 percent of Americans resided in 1928, had poor or no service—the entire state of South Carolina had only two radio stations (Craig 2004a).

6. During the Great Migration between 1910 and 1970, about six million Blacks left the South (US Census Bureau 2012). In the decade discussed here, 1930–1940, about 400,000 Blacks left the South (Gregory 2005). Migrants had increased access to public spaces, including ones that provided access to radio (see Grossman 1991 for an in-depth examination of increased access to public spaces in Chicago).

7. The corresponding figures expressed in inflation-adjusted 2022 dollars are $33,915 and $1,343, respectively.

8. Listeners of such "crystal sets," which were batteryless, had to use headsets (Vaillant 2002a, 62).

9. Similarly, in Nazi-occupied France, radio was the last thing that the French sold for food and other necessities of life (Vaillant 2017).

10. The farmers' efforts to acquire receiver sets were aided by the industry, which offered installment plans (Compaine 2001a).

11. Moreover, radios could be accessed in the homes of friends and relatives or in public places (Craig 2004a). This further reduced pressure for public policy intervention.

12. It has now spread to other countries, including the UK and Canada (Ali 2016).

13. The Radio Act of 1927, Public Law No. 632, February 23, 1927, 69th Congress, https://www.americanradiohistory.com/Archive-FCC/Federal%20Radio%20Act%201927.pdf (accessed July 28, 2012).

14. March 28, 1928, 70th Congress, S. 2317, Public Law No. 195, Section 5 (amendment to the second paragraph of section 9 of the Radio Act of 1927), http://www.loc.gov/law/help/statutes-at-large/70th-congress/session-1/c70s1ch263.pdf (accessed July 28, 2019).

15. Interestingly, in the then–newly formed Soviet Union, we find an edifying contrast. Vladimir Lenin, who pushed his scientists and engineers to develop radio, wrote to his point person in 1920: "The newspaper without paper and 'without distances' which you are creating will be a great thing. I promise to give you any and every assistance in this and similar projects" (Guback and Hill 1972, 11–12). He wanted a powerful transmitter in Moscow that could serve a radius of 1,300 miles. In a 1921 note to a senior official, he envisioned that over a powerful transmitter, "all Russia will hear a newspaper read in Moscow" (Guback and Hill 1972, 16).

16. They included companies such as RCA, Westinghouse, AT&T, and General Electric (GE).

17. In particular, for those invested in local broadcasting, the very notion of national stations was anathema (Messere 2005).

18. The Communications Act of 1934 created the FCC, which replaced the FRC for radio regulation, and took over the regulation of wire communications from the Interstate Commerce Commission.

19. In the first instance of network broadcasting, on October 28, 1922, Westinghouse and AT&T used long-distance telephone lines to broadcast a University of Chicago–Princeton University football game in Chicago to other parts of the country (Banning 1946; Socolow 2001). In January 1923, WNAC in Boston transmitted a five-minute saxophone solo over telephone lines to WEAF in New York. A few months later, WEAF and WMAF in Dartmouth, Massachusetts, established a permanent (i.e., nonoccasional) network (Sterling and Kittross 1990). NBC and CBS were established in 1926 and 1927, respectively (Craig 2010a). In 1927, the two networks together had 6.4 percent of stations. Four years later, the corresponding figure was 30 percent (McChesney 1993). In effect, network broadcasting was building up when the Radio Act of 1927 was passed. But its true significance was not understood at that point in time.

20. From an advertiser's standpoint, a network offered great efficiencies. Instead of contracting with myriad radio stations and then monitoring whether the advertisements were actually run in all the different markets, all the advertiser had to do was to contract with a network (Horwitz 1989; Kirkpatrick 2006; Socolow 2001; Streeter 1996).

21. In the networked world, localism has become decoupled from locality (i.e., it is no longer anchored in geography). In the case of news, Ali (2016) asks: "Is local news, for instance, specific to an immediate geographic locality? Is it news that is of interest to the local population? Or is it news produced within the locality? In other words, what should 'count' as local?" (107). In this context, the debate has

pitted "spatial localism" against "social localism," with the former centered on a place and the latter on a commonality of interests—which is not necessarily limited to the people of a locality. For instance, aboriginal producers want to be thought of as "local broadcasters" even though their networks span thousands of miles (Ali 2016). As Jean LaRose, the chief executive officer of the Aboriginal Peoples Television Network (APTN) in Canada, explained to her nation's parliament: "This is a different way of looking at local programming. Programming that reflects Nunavut and Nunavik is local, from our point of view, even though the communities it serves are spread out over a region that represents a large percentage of Canada's land mass" (Ali 2016, 121).

22. On the fourth anniversary of KDKA's broadcast of the Harding-Cox election results, the *New York Times* (1924b, 6) wrote: "Today, when radio broadcasting is so much a part of our national life, the crude methods of four years ago and its limited application seem to belong to another century." In a nutshell, radio became part of the national scene in three to four years.

23. Educators founded the Association of College and University Broadcasting Stations (ACUBS) in 1925, saying in its constitution:

Believing that radio is in its very nature one of the most important factors in our national and international welfare, we, the representatives of institutions of higher learning, engaged in educational broadcasting, do associate ourselves together to promote, by mutual cooperation and united effort, the dissemination of knowledge to the end that both the technical and educational feature of broadcasting may be extended to all. (Saettler 1990, 216)

The ACUBS later became the National Association of Educational Broadcasters (NAEB), a forebear of the Public Broadcasting Service (PBS) and National Public Radio (NPR).

24. First, the FCC put in place programming guidelines that specified that renewal of licenses of stations with less than 5 percent of airtime devoted to local programming would be subject to review. In 1984, it withdrew these guidelines because they turned out to be ineffective. In the same vein, the FCC first stipulated that cable operators with more than 3,500 subscribers could carry TV programs only if they had local facilities for program production. In 1974, it rescinded this requirement because the local origination programming so produced had an extremely small audience, saying that "mandatory origination scheme is not likely to be the most effective means of fostering local expression programming" (Napoli 2001, 212). But then, a little over a decade later, it instituted Main Studio and Program Origination Rules, which required radio and TV stations to (1) maintain a studio, with full-time management and staff, in the principal community of their licensed service area; and (2) originate more than 50 percent of their nonnetwork programs from the said studio or nearby ancillary facilities (Braman 2007; Napoli 2001). Once again, the FCC concluded that this requirement was not productive. In 2017, it eliminated the principal community studio requirement. Now, as noted earlier, broadcasters basically only need to have a local or toll-free number (Feldman 2017).

25. Jerome Barron (1967), who provides an intellectual articulation of such bottom-up efforts, argues that the First Amendment should not apply only to utterances already made, as this creates an asymmetry by protecting the speech of those who

can speak, but without giving voice to those lacking resources. He holds that the right to access is implicit in the First Amendment, and public policy should provide the means for its exercise. Further, while the utterance-oriented interpretation of the First Amendment was understandable when it was ratified, given that newspapers then were small, local, and accessible to the citizens, with regard to our mediatized world, writing in the pre-internet era, Barron asks: What about those who do not get an opportunity to speak via the media at a time when standing on a soapbox to give a speech is no longer effective for public discussion?

26. Further, unlike public television, where financial contributions are earnestly sought, there is no such need for public access television because it is funded directly as part of franchise terms.

27. Interestingly, since the turn of the century, low-power FM (LPFM) activists have sought to provide a corrective. LPFM was a pirate activity until 2000, when the FCC established it as a new class of radio station. Now, the US has about 2,200 LPFM stations (National Archives and Records Administration 2020). They are limited to noncommercial broadcasting and 100 watts of power—giving a service range of about a 3.5-mile radius (FCC n.d.). The activists see LPFM, a community-scale technology, as a vehicle for cultivating the particularities of the local, in marked contrast to the global-scale internet (Dunbar-Hester 2014).

28. Even corporate executives believed that radio advertisements would be too intrusive because, unlike newspaper and magazine advertisements, they could not be avoided. In effect, a stranger's voice would be coming into a house uninvited and making a sales pitch.

29. He was the founder of Henry Field Seed Company, a catalog mail-order business, and the star on-air personality on the broadcasts of KFNF, the radio station he set-up on the top of his premises.

30. The success of KFNK inspired other Iowa stations to follow suit. Iowa, with the nation's principal direct selling stations, itself became the center of controversy (St. Austell 1928).

31. It included a daily news digest prepared by the US Department of Agriculture (USDA), which since the early 1920s has promoted radio ownership by farmers and worked with broadcasters to provide agriculture-oriented, specialized programs. By 1928, USDA's radio service was producing its own programs in script form, which it provided to interested broadcasters free of charge (Craig 2009a, 2009b, 2010a).

32. In 1924, Sears, Roebuck & Co. established WLS (which stood for "World's Largest Store"), which launched the careers of music and comedy superstars such as Bradley Kincaid, Gene Autry, Pat Buttram, George Gobel, and Patsy Montana (Barron 1997; PBS n.d.).

33. In 1922, *Country Life* magazine carried an article entitled "Removing the Last Objection to Living in the Country." The "last objection" here was a stand-in for rural isolation, whose removal would be the grant of a new technology that was an "adaptation of the wireless to a form in which anybody . . . can use the wireless and at a price within the reach of even the most modest purse" (*Country Life* 1922, 63).

34. There were also many small and telling changes. For instance, instead of being addressed as "Ladies and Gentlemen," the members of the audience were addressed as "Friends" (Kirkpatrick 2006).

35. The *National Barn Dance,* discussed earlier, was also trans-local.

36. In fact, many local stations broadcast church services for shut-ins (Vaillant 2002a).

37. The 1920 census showed that more than half of the US population (51.2 percent) was living in urban areas at that time.

38. Conversely, if the system that thus emerged did not accord with the interests of the powers that be, the zero-tolerance policy for interference would not have had the centrality that it had in the formation of the American broadcasting system. Other organizing principles would have been in play as well.

39. Local radio stations often voluntarily organized weekly "silent nights," when they all refrained from broadcasting so that their listeners could pick up faint signals from distant places (Kirkpatrick 2011, 256).

40. We would like to thank Kevin Howley for an edifying discussion on the genesis of the present-day German broadcasting system.

41. Initially, policymakers relied on telegraph analogs and precedents for their decision-making on radio-related issues. With the emergence of broadcasting, these policymakers needed a new approach, but they struggled because they could not find a good analog for it. Eventually, they settle on the PICON ("public interest, convenience and necessity") standard, which had long been used for the regulation of common carriers and public utilities, because it was broad and flexible (Krasnow and Goodman 1998; Sawhney et al. 2010). The FRC cast this broad and flexible (but also vague) language in ways that furthered the advancement of networks over independent local radio stations (see Vaillant 2002a for specific examples).

42. Many nonprofits could not even afford the expense of defending their licenses before the FRC in Washington, DC, which they had to do every three months. Commercial broadcasters frequently challenged their licenses—in any month, as many as half of the nonprofits saw their licenses challenged (McChesney 1993).

43. General Order 32, issued in May 1928.

44. After the passage of the Radio Act of 1912, which established the principle of legally sanctioned exclusive spectrum allocation, the US Department of Commerce allocated what were then understood to be useful parts of the spectrum to the US navy (600–1,600-meter range) and commercial operators (the 200–600-meter range and above 1,600 meters). The amateurs were allocated the potion of the spectrum below 200 meters, which was thought of as a wasteland. They went on to develop shortwave technologies, which proved to be far superior to the ones that the industry was experimenting with for this portion of the spectrum (Streeter 1996). This was a lasting contribution, as shortwave has long been the mainstay of international broadcasting by the British Broadcasting Corporation (BBC) and others.

8 THE INTERNET

1. Stalwarts include Vint Cerf and Bob Kahn ("fathers of the internet") and others involved with the Internet Engineering Task Force, which developed internet standards to "facilitate, support, and promote the evolution and growth of the internet as a global research communications infrastructure" (Cerf, Kahn, and Chapin 1992, n.p.).

2. The official history also highlights that early foundational research took place concurrently at the Massachusetts Institute of Technology (1961–1967), the RAND Corporation (1962–1965), and the UK's National Physical Laboratory (1964–1967), without the researchers knowing about developments at other institutions.

3. While the Clinton-Gore administration (1993–2001) is generally associated with the commercialization of the internet, policymakers were deliberating on such a move in the late 1980s. For instance, the Office of Technology Assessment's 1989 background paper *High Performance Computing & Networking for Science* asks:

One of the key issues centers around the extent to which deliberate creation of a market should be built into network policy, and into the surrounding science policy system. There are those who believe that it is important that the delivery of network access and services to academics eventually become a commercial operation, and the current Federal subsidy and apparently "free" services will get academics so used to free services that there will never be a market. How do you gradually create an information market, for networks, or for network accessible value-added services? (Office of Technology Assessment 1989, 32)

4. See chapter 7, on broadcasting, for a discussion on how Compaine draws on the experience with radio to make his case.

5. Google, Google Map Maker has closed, https://support.google.com/mapmaker /answer/7195127?hl=en (accessed April 25, 2022).

6. United Airlines, Taxes and Fees, https://www.united.com/ual/en/us/fly/booking /flight/taxes.html#:~:text=A%20fee%20will%20be%20charged,in%20person%20 at%20the%20airport (accessed July 12, 2020).

7. To familiarize himself with *Fortnite*, a video game that his 11-year-old daughter was drawn to, Tom Vanderbilt started playing it and found her engaging in "intense negotiations with her largely male teammates . . . working in tandem to devise strategies, tactfully soliciting input or advancing her own opinion, deftly delegating responsibilities. At times it seemed less like a game than a virtual workplace" (Vanderbilt 2020, 32).

8. To explore this opacity, Ekbia (2018) introduces the notion of "the alter-sphere," where the individual, unaware of the rules of the game, "is objectified in the schemes of other actors," often unknown ones (77–78).

9. We have been through such cycles many times before. Standage makes this point sharply by calling the telegraph, an obsolete technology, as the internet, our cynosure, of its times—that is, the Victorian internet. Taking the long view, Standage (1999) observes: "The optimistic claims now being made about the internet are merely the recent examples in a tradition of technological utopianism that goes back to the first transatlantic telegraph cables, 150 years ago" (211). Approaching from the other direction, from the oldest to the newest, Mosco (2004) makes the same point: "All the

wonders that were forecast for the telegraph, electricity, the telephone, and broadcasting were invested in the computer" (2).

10. Digital Detox, Camp Grounded, Summer Camp for Adults, https://www.digital detox.com/experiences/camp-grounded (accessed July 14, 2020).

11. The public and policymakers celebrate technology, dazzled by its benefits—the obvious ones, at any rate. Scholars, in the technopessimist tradition, provide correctives, spotlighting negative consequences—subtle ones. In the 1950s Jacques Ellul (1964) warned against our single-minded pursuit of efficiency, something that we take to be inherently good, saying: "technique transforms everything it touches into a machine" (4). He meant this both literally (physical machinery) and metaphorically (social machinery)—that is, highly rationalized human interactions. More recently, Dreyfus (2009) provided a corrective to John Perry Barlow's (1996) celebrated view of cyberspace—"a world that is both everywhere and nowhere, but it is not where bodies live" (n.p.). Through a critique of distance learning, telepresence, and online anonymity, he spotlighted the negative consequences of disembodiment. He expressed special concern about the internet, which "affects people in ways that are different from the way most tools do because it can become the main way its users relate to the rest of the world" (Dreyfus 2009, 3). Other scholars, technorealists, argue that we need to go beyond the optimist/pessimist binary and understand technologies in their full complexity, both welcoming and fearing them (for an overview of technorealist principles and values, see Shenk, Shapiro, and Johnson 1998). Of the works in this tradition, Matthew Fuller and Andrew Goffey's *Evil Media* is noteworthy from the standpoint of our project. They argue that "conventional media studies . . . are far too concerned with thinking things through from the spectator's perspective" (Fuller and Goffey 2012, 2). They call for the study of the less visible "gray media"—databases, software for teamwork, project-planning techniques and tools, etc. Fuller and Goffey decenter the spectator's perspective and open up a new way of looking at media. They make this move once. We, on the other hand, make many such moves, and, moreover, in a systematic manner. With regard to connectivity in particular, scholars such as Innis (1950, 1951), Carey (1989), Samarajiva and Shields (1990a, 1990b), and Schivelbusch (2014) have provided incisive critiques. They alert us to the negative consequences that we are ordinarily unable to decipher. However, they do not give us guidance on how to navigate the complexities of connectivity, except for Samarajiva and Shields (1990b). In the context of telecommunications in developing countries, Samarajiva and Shields suggest that rural communities should first develop lateral links with other nearby communities before connecting with a big city. In effect, they suggest an unorthodox sequencing of links—not the nature of connectivity itself. We call for a rethinking of the nature of connectivity. Also, our analysis is not limited to one particular context.

12. At first, the measure of internet access was occasional use. Critics, arguing that it underplayed the severity of the digital divide, pushed for a new measure—frequency of use. This shift was consequential. For instance, in 2006, the proportion of population online was 73 percent, on the basis of the occasional use measure, or 48 percent, on the basis of the daily use measure (Mossberger et al. 2008).

9 CONCLUSION

1. We use the word "discoursed" because many points were not debated per se (i.e., considered point-counterpoint). Different stakeholders made different points in different forums, as opposed to counterposing them on shared forums, as in a debate. For our part, we searched for different points made by different stakeholders in different forums as exhaustively as we could—constructing an inventory of points made by different stakeholders.

2. More recently, the scope of the problem is taken as not only securing access to technologies, but also ensuring the requisite skills to use them.

3. This is not to deny the transformative power of the link, but to underscore its limits. It is not so transformative as to level the power asymmetries between the nodes that it connects. At the same time, it does change the arena in which the power asymmetries between the nodes play out, and on that score, it is transformative. All in all, connectivity redesigns the arena—it does not equalize the participants.

4. REGULATEL (Foro Latinoamericano de Entes Reguladores de Telecomunicaciones) stands for Latin American Forum of Telecommunications Regulatory Entities.

REFERENCES

Aberdeen Daily News (1915). Motorization of mail work. *Aberdeen Daily News*, October 11, 1.

Adams, F. (1875). *The free school system of the United States.* London: Chapman & Hall.

Agre, P. (1997). *Computation and human experience.* Cambridge: Cambridge University Press.

Akhtar, A., and M. Ward (2020). Bill Gates and Steve Jobs raised their kids with limited tech—and it should have been a red flag about our own smartphone use. *Business Insider*, May 15. Retrieved on July 14, 2020, from https://www.businessinsider.com/screen-time-limits-bill-gates-steve-jobs-red-flag-2017-10.

Alglave, E., and J. Boulard (1884). *The electric light: Its history, production, and applications.* Translated from French by T. O'Conor Sloane. New York: D. Appleton.

Ali, C. (2016). Critical regionalism and the policies of place: Revisiting localism for the digital age. *Communications Theory* 26(2): 106–127.

Ali, C. (2017). *Media localism: The policies of place.* Urbana: University of Illinois Press.

Ali, C. (2020). The other homework gap: Post-secondary education during COVID-19. Benton.org, April 7. Retrieved on July 11, 2020, from https://www.benton.org/blog/other-homework-gap-post-secondary-education-during-covid-19.

Alexander K., and K. F. Jordan (1973). *Legal aspects of educational choice: Compulsory attendance and student assignment.* Topeka, KS: National Organization of Legal Problems in Education.

Anderson, F. (1963). *Library program, 1911–1961.* New York: Carnegie Corporation of New York.

Arakpogun, E. O., R. Wanjiru, and J. Whalley (2017). Impediments to the implementation of universal service funds in Africa: A cross-country comparative analysis. *Telecommunications Policy* 41(7–8): 617–630.

Arboleda, J. (1981). *Open admissions to higher education and the life chances of lower class students: A case study from Columbia.* Doctoral dissertation, Indiana University, Bloomington.

Atwood, R. A. (1984). *Telephony and its cultural meanings in southeastern Iowa, 1900–1917.* Doctoral dissertation, University of Iowa, Iowa City.

Atwood, A. (1986). Routes of rural discontent: Cultural contradictions of Rural Free Delivery in southeastern Iowa, 1899–1917. *Annals of Iowa* 48(5): 264–273.

Bailey, K. (1987). Soviet sponsor spread of AIDS disinformation. *Los Angeles Times*, April 19. Retrieved on July 7, 2020, from https://www.latimes.com/archives/la-xpm -1987-04-19-op-1947-story.html.

Bakke, G. (2016). *The grid: The fraying wires between Americans and our energy future.* New York: Bloomsbury.

Banning, W. P. (1946). *Commercial broadcasting pioneer: The WEAF experiment, 1922–1926.* Cambridge, MA: Harvard University Press.

Barfield, R. (1996). *Listening to radio, 1920–1950.* Westport, CT: Praeger.

Barlow, J. P. (1996). A declaration of the independence of cyberspace. Retrieved on July 17, 2020, from eff.org/cyberspace-independence.

Barnett, W. P., and G. R. Carroll (1993). How institutional constraints affected the organization of early U.S. telephony. *Journal of Law, Economics and Organization* 9(1): 98–126.

Barnouw, E. (1966). *A tower in Babel: A history of broadcasting in the United States to 1933, Volume 1.* New York: Oxford University Press.

Barron, H. S. (1997). *Mixed harvest: The second great transformation in the rural North, 1870–1930.* Chapel Hill: University of North Carolina Press.

Barron, J. A. (1967). Access to the press: A new first amendment right. *Harvard Law Review* 80(8): 1641–1678.

Bartels, M. (2019). NASA will need your help mapping asteroid Bennu. Space.com, March 7. Retrieved on July 13, 2020, from https://www.space.com/nasa-needs-help -mapping-asteroid-bennu.html

Behringer, W. (2006). Communications revolutions: A historiographical concept. *German History* 24(3): 333–374.

Belinfante, A. (2006). *Telephone penetration by income by state.* Washington, DC: Federal Communications Commission.

Bell, D. (1980). Introduction. In *The computerization of society: A report to the president of France*, S. Nora and A. Minc, vii–xvi. Cambridge, MA: MIT Press.

Beniger, J. R. (1986). *The control revolution: Technological and economic origins of the information society*. Cambridge, MA: Harvard University Press.

Benkler, Y. (2006). *The wealth of networks: How social production transforms markets and freedom*. New Haven, CT: Yale University Press.

Bilton, N. (2014). Steve Jobs was a low-tech parent. *New York Times*, September 10. Retrieved on July 14, 2020, from https://www.nytimes.com/2014/09/11/fashion /steve-jobs-apple-was-a-low-tech-parent.html?_r=0.

Binder, F. M. (1974). *The age of the common school, 1830–1865*. New York: John Wiley & Sons.

Birenbaum, W. M. (1971). Something for everybody is not enough. In *Higher education for everybody?* ed. W. T. Furniss, 65–82. Washington, DC: American Council on Education.

Black, M. (1962). *Models and metaphors*. Ithaca, NY: Cornell University Press.

Blalock, H. W. (1940). Streamlining rural telephone service. *Public Utilities Fortnightly*, October 10, 466–473.

Boone, R. G. (1889). *Education in the United States: Its history from the earliest settlements*. New York: D. Appleton.

Boston Public Library (1878). *Proceedings at the dedication of the Jamaica Plain branch of the Boston Public Library*. Boston: Boston Public Library.

Bowles, F. (1966). Observations and comments. In *Universal higher education*, ed. E. J. McGrath, 235–245. New York: McGraw-Hill.

Bradshaw, T. (2018). Snap's chief Evan Spiegel: Taming tech and fighting with Facebook. *Financial Times*, December 28. Retrieved on July 14, 2020, from https://www.ft .com/content/fdfe58ec-03a7-11e9-9d01-cd4d49afbbe3.

Braman, S. (2006). *Change of state: Information, policy, and power*. Cambridge, MA: MIT Press.

Braman, S. (2007). The ideal v. the real in media localism: Regulatory implications. *Communication Law and Policy* 12(3): 231–278.

Brock, G. W. (1981). *The telecommunications industry: The dynamics of market structure*. Cambridge, MA: Harvard University Press.

Brooks, J. (1976). *Telephone: The first hundred years*. New York: Harper & Row.

Brown, D. C. (1970). *Rural electrification in the South: 1920–1955*. Doctoral dissertation, University of California, Los Angeles.

Brown, D. C. (1980). *Electricity for rural America: The fight for the REA*. Westport, CT: Greenwood Press.

Browning, J. (1996). Universal service (an idea whose time is past). *Wired*, September. Retrieved on May 25, 2018, from https://www.wired.com/1994/09/universal-access/.

Brownson, O. (1839). Second annual report of the board of education, together with the second annual report of the secretary of the board *Boston Quarterly Review* 2(4): 393–434.

Bucy, E. P. (2000). Social access to the internet. *Harvard International Journal of Press/ Politics* 5(1): 50–61.

Bureau of Labor Statistics (2012). Consumer expenditure surveys: Spending on newspapers and magazines slides as spending for internet access soars. Retrieved on July 1, 2019, from https://www.bls.gov/cex/newspapers.htm.

Bureau of Labor Statistics (2019). Prices and spending: Are most Americans cutting the cord on landlines? Retrieved on July 1, 2019, from https://www.bls.gov/opub /btn/volume-8/are-most-americans-cutting-the-cord-on-landlines.htm.

Bureau of the Census (1976). *Historical statistics of the United States, colonial times to 1970.* Washington, DC: United States Government Printing Office.

Burgess, J. T. F. (2013). *Virtue ethics and the narrative identity of American librarianship 1876 to present.* Doctoral dissertation, University of Alabama, Tuscaloosa.

Burritt, M. C. (1931). Selling farmers electricity on community basis. *Electrical World*, October 17, 696–698.

Button, H. W., and E. F. Provenzo, Jr. (1989). *History of education and culture in America.* Englewood Cliffs, NJ: Prentice-Hall.

Butts, R. F. (1978). *Public education in the United States.* New York: Holt, Rinehart and Winston.

Canning, C. M. (2005). *The most American thing in America: Circuit Chautauqua as performance.* Iowa City: University of Iowa Press.

Carey, J. (1989). *Communication as culture.* Boston: Unwin Hayman.

Carey, J. W., and J. T. Quirk (1970). The mythos of the electronic revolution. *The American Scholar* 39(3): 395–424.

Carlton, D. W., and J. M. Klamer (1983). The need for coordination among firms, with special reference to network industries. *University of Chicago Law Review* 50: 446–465.

Carlton, F. T. (1906). *Economic influences upon educational progress in the United States, 1820–1850.* Doctoral dissertation, University of Wisconsin, Madison.

Carmody, J. M. (1939). Rural electrification in the U.S. *Annals of the American Academy of Political and Social Science* 201: 82–95.

Castells, M. (2011). *The rise of the network society.* New York: Wiley.

Cerf, V., B. Kahn, and L. Chapin (1992). Announcing the Internet Society. Retrieved on July 22, 2020, from https://www.internetsociety.org/internet/history-of-the-internet /announcing-internet-society/.

Chicago Daily Tribune (1900a). Good roads and Rural Free Delivery. *Chicago Daily Tribune*, December 17, 6.

Chicago Daily Tribune (1900b). New plan for Rural Free Delivery. *Chicago Daily Tribune*, March 5, 6.

Chicago Daily Tribune (1904). To impair Rural Free Delivery. *Chicago Daily Tribune*, March 2, 6.

Chicago Daily Tribune (1905). Rural Free Delivery. *Chicago Daily Tribune*, December 8, 8.

Childs, M. (1974). *The farmer takes a hand: The electric power revolution in rural America*. New York: Da Capo Press.

Chilton, O. (2012). *The office building in the future*. Mulgrave, Australia: Images Publishing Group.

Christie, J. (1983). *Morris Llewellyn Cooke: Progressive engineer*. New York: Garland Publishing.

Cincinnati Enquirer (1905). Little corner for little people. *Cincinnati Enquirer*, July 23, D7.

Cincinnati Enquirer (1907). Rural Free Delivery. *Cincinnati Enquirer*, December 12, 6.

Cochran, A. B. III (1995). We participate, they decide: The real stakes in revising Section 8(a)(2) of the National Labor Relations Act. *Berkeley Journal of Employment and Labor Law* 16(2): 458–519.

Cohen, N. (2011). Define gender gap? Look up Wikipedia's contributor list. *New York Times*, January 30. Retrieved on June 28, 2019, from https://www.nytimes.com/2011/01/31/business/media/31link.html.

Cohn, J. A. (2017). *The grid: Biography of an American technology*. Cambridge, MA: MIT Press.

Cole, D. (1936). *Electric power on the farm*. Washington, DC: United States Government Printing Office.

Cole, H., and P. Murck (2007). The myth of the localism mandate: A historical survey of how the FCC's actions belie the existence of a governmental obligation to provide local programming. *CommLaw Conspectus* 15(2): 339–371.

Colman's Rural World (1900). The farmer's influence: The grout bill, Rural Free Delivery. *Colman's Rural World*, December 12, 1.

Columbus Ledger (1905). Kick is made of numbering boxes. *Columbus Ledger*, September 28, 7.

Compaine, B. (1986). Information gaps: Myth or reality? *Telecommunications Policy* 10(1): 5–12.

Compaine, B. (2001a). Information gaps: Myth or reality? In *The digital divide: Facing a crisis or creating a myth?* ed. B. Compaine, 105–118. Cambridge, MA: MIT Press.

Compaine, B. (2001b). Introduction to the section "The context: Background and texture." In *The digital divide: Facing a crisis or creating a myth?* ed. B. Compaine, 101–103. Cambridge, MA: MIT Press.

Conaty, P., and D. Bollier (2014). Toward an open co-operativism: A new social economy based on open platforms, co-operative models and the commons. Report on a Commons Strategies Group Workshop, Berlin, August. Retrieved on January 6, 2021, from https://base.socioeco.org/docs/open_20co-operativism_20report_2c_20january_202015_0.pdf.

Constitution of the State of Indiana (1816). Article IX, Section 2. Retrieved on February 29, 2021, from https://www.in.gov/history/about-indiana-history-and-trivia/explore-indiana-history-by-topic/indiana-documents-leading-to-statehood/constitution-of-1816/article-ix/.

Constitution of the United States (1788). Retrieved on April 22, 2022, from https://www.archives.gov/founding-docs/constitution-transcript

Cooke, M. L. (1935). The new viewpoint. *Rural Electrification News*, October, 1–4.

Cooper, T. E. (1940). Rural electrification and the country doctor. *Rural Electrification News*, March, 12.

Country Life (1922). Removing the last objection to living in the country. *Country Life*, February, 63.

Counts, G. S. (1967). Foreword. In *Presidential statements on education: Excerpts from inaugural and state of the union messages, 1789–1967*, ed. M. J. Thomas, ix–x. Pittsburgh, PA: University of Pittsburgh Press.

Coutard, O., R. Hanley, and R. Zimmerman (eds.) (2004). *Sustaining urban networks: The social diffusion of large technical Systems*. London: Routledge.

Cowles, J. L. (2012). A parcels post: A cent a pound. In *The American postal network, 1792–1914, Vol. 4*, ed. R. R. John, 229–237. London: Pickering & Chatto.

Coyle, D. C. (1936). *Electric power on the farm: The story of electricity, its usefulness on farms, and the movement to electrify rural America*. Washington, DC: Rural Electrification Administration.

Craig, S. (2004a). How America adopted radio: Demographic differences in set ownership reported in the 1930–1950 U.S. censuses. *Journal of Broadcasting & Electronic Media* 48(2): 179–195.

Craig, S. (2004b). *Mail order radios: Receiver evolution in the pages of the Sears, Roebuck and Montgomery Ward catalogs, 1920–1950*. Paper presented at the Popular Culture Association Convention, San Antonio, Texas, April.

Craig, S. (2009a). *Framers and radio during the Great Depression*. Paper presented at the Annual Meeting of the Agricultural Historical Society, Little Rock, Arkansas, June.

Craig, S. (2009b). *Out of the dark: A history of radio and rural America*. Tuscaloosa: University of Alabama Press.

Craig, S. (2010a). Daniel Starch's 1928 survey: A first glimpse of the U.S. radio audience. *Journal of Radio & Audio Media* 17(2): 182–194.

Craig, S. (2010b). *Rural isolation and the arrival of radio: Conceptualizing the meaning of social contact.* Paper presented at the Annual Meeting of the Western Social Science Association, Reno, Nevada, April.

Cremin, L.A. (1980). *American education: The national experience, 1783–1876.* New York: Harper & Row.

Curtis, S., W. Gesler, G. Smith, and S. Washburn (2000). Approaches to sampling and case selection in qualitative research: Examples in the geography of health. *Social Science & Medicine* 50: 1001–1014.

Cushing, M. (1893). *The story of our post office: The greatest government department in all its phases.* Boston: A. M. Thayer.

Daily Picayune (1908). Rural mail service. *Daily Picayune*, December 1, 8.

Dallas Morning News (1902). Department gives warning. *Dallas Morning News*, December 27, 12.

Davidson, J. (2018). Thieves targeted $12 billion through IRS tax fraud. *Washington Post*, October 19. Retrieved on July 6, 2020, from https://www.washingtonpost.com /politics/2018/10/19/thieves-targeted-billion-through-irs-tax-fraud.

Davis, H. (1930). American beginnings. In *Radio and its future*, ed. M. Codel, 3–11. New York: Harper.

Dawson, J. A. (ed.). (1945). *The travail of Russia: Human travail of Russia.* Melbourne, Australia: Workers' Literature Bureau.

Denzin, N. K. (2001). *Interpretive interactionism.* 2nd ed. Thousand Oaks, CA: SAGE.

DeSilver, D. (2020). Before the coronavirus, telework was an optional benefit, mostly for the affluent few. PewResearch.com, March 20. Retrieved on July 11, 2020, from https://www.pewresearch.org/fact-tank/2020/03/20/before-the-coronavirus-tele work-was-an-optional-benefit-mostly-for-the-affluent-few/.

Deutsch, J. (1944). *Rural electrification in North Carolina.* Master's thesis, University of North Carolina, Chapel Hill.

Dholakia, R. R. (2006). Gender and IT in the household: Evolving patterns of internet use in the United States. *The Information Society* 22(4): 231–240.

Didion, J. (1990). *Slouching towards Bethlehem: Essays by Joan Didion.* New York: Farrar, Straus and Giroux.

Dieken, G. (1936). Rural electrification training course given to Iowa farm agents. *Rural Electrification News*, November, 18–19.

Dilts, M. M. (1941). *The telephone in a changing world.* New York: Longmans, Green & Co.

Ditzion, S. (1947). *Arsenals of a democratic culture: A social history of the American public library movement in New England and the Middle States, from 1850 to 1900*. Chicago: American Library Association.

Dohrn-van Rossum, G. (1996). *History of the hour: Clocks and modern temporal orders*. Chicago: University of Chicago Press.

Donnelly, F. P. (2015). Regional variations in average distance to public libraries in the United States. *Library & Information Science Research* 37(4): 280–289.

Douglas, G. (1987a). *The early days of radio broadcasting*. Jefferson, NC: McFarland.

Douglas, M. (1966). *Purity and danger: An analysis of concepts of pollution and taboo*. London: Routledge.

Douglas, S. (1986). Amateur operators and American broadcasting: Shaping the future of radio. In *Imagining tomorrow: History, technology, and the American future*, ed. J. Corn, 35–57. Cambridge, MA: MIT Press.

Douglas, S. (1987b). *Inventing America broadcasting, 1899–1922*. Baltimore: Johns Hopkins University Press.

Dreyfus, H. L. (2009). *On the internet*. Abingdon, UK: Routledge.

Dunbar-Hester, C. (2014). *Low power to the people: Pirates, protest, and politics in FM radio activism*. Cambridge, MA: MIT Press.

Dyer-Witheford, N. (1999). *Cyber-Marx: Cycles and circuits of struggle in high-technology capitalism*. Urbana: University of Illinois Press.

Educational Policies Commission (1964). *Universal opportunity for education beyond the high school*. Washington, DC: National Education Association.

Edwards, N., and H. G. Richey. (1963). *The school in the American social order*. Boston: Houghton Mifflin.

Edwards, P. N. (2010). *A vast machine: Computer models, climate data, and the politics of global warming*. Cambridge, MA: MIT Press.

Eisenhardt, K. M. (1989). Building theories from case study research. *Academy of Management Review* 14(4): 532–550.

Ekbia, H. (2016). Digital inclusion and social exclusion: The political economy of value in a networked world. *The Information Society* 32(3): 165–75.

Ekbia, H. (2018). The tyranny of the alter-sphere. *Figurationen* 19(1): 71–88.

Electrical World (1931). Declares rural lines cost too much. *Electrical World*, November 28, 952–953.

Electrical World (1935). Untitled. *Electrical World*, January 5, 56.

Ellul, J. (1964). *The technological society*. New York: Alfred A. Knopf.

Engelman, R. (1990). The origins of public access cable television 1966–1972. *Journalism Monographs* 123: 1–47.

Erdman, H. E. (1930). Some social and economic aspects of rural electrification. *Journal of Farm Economics* 12(2): 311–19.

Esping-Andersen, G. (1998). *The three worlds of welfare capitalism*. Princeton, NJ: Princeton University Press.

Etzioni, A. (2011). Cybersecurity in the private sector. *Issues in Science and Technology*, 28(1). Retrieved on July 7, from https://issues.org/etzioni-2/.

Everett, E. (1851). A public library (Extract from a letter to the Mayor of Boston, dated June 7, 1851). *The Massachusetts Teacher* 4(8): 255–256.

Fairlie, R. W. (2014). *Race and the digital divide*. UC Santa Cruz Working Paper Series. Santa Cruz: Department of Economics, University of California, Santa Cruz. Retrieved on January 6, 2021, from https://escholarship.org/content/qt48h8h99w/qt48h8h99w.pdf.

Federal Bureau of Investigation (FBI) (2020). *2019 Internet crime report*. Washington, DC: Internet Crime Complaint Center, Federal Bureau of Investigation.

Federal Communications Commission (FCC) (n. d). Low-power FM (LPFM) broadcast radio stations. Retrieved on July 30, 2020, from https://www.fcc.gov/media/radio/lpfm.

Federal Communications Commission (FCC) (1998). *Telephone subscribership in the United States*. Washington DC: Federal Communications Commission. Retrieved on July 4, 2021, from https://transition.fcc.gov/Bureaus/Common_Carrier/Reports/FCC-State_Link/IAD/subs0798.pdf.

Feldman, P. J. (2017). FCC eliminates broadcast main studio rules, related staffing, and program origination requirements; controversial order passes three-two along party lines. *CommLawBlog*, October 25. Retrieved on June 15, 2018, from https://www.commlawblog.com/2017/10/articles/fcc/fcc-eliminates-broadcast-main-studio-rules-related-staffing-and-program-origination-requirements-controversial-order-passes-three-two-along-party-lines/.

First Assistant Postmaster-General (1893). Report of the first assistant postmaster-general for the fiscal year ending June 30, 1893. In *Report of the Postmaster-General of the United States; being part of the message and documents communicated to the two houses of Congress at the beginning of the second session of the fifty-third Congress, 29–131*. Washington, DC: Government Printing Office.

Fischer, C. S. (1987). The revolution in rural telephony, 1900–1920. *Journal of Social History* 21(1): 5–26.

Fischer, C. S. (1992). *America calling: A social history of the telephone to 1940*. Berkeley: University of California Press.

Fletcher, W. I. (1904). *Public libraries in America*. Boston: Roberts Brothers.

Flichy, P. (2007). *The internet imaginaire*. Translated by L. Carey-Libbrecht. Cambridge, MA: MIT Press.

Foucault, M. (1995). *Discipline and punish*. Translated by Alan Sheridan. New York: Vintage Books.

Fox, S. (2015). From nurses to social workers, see how public libraries are serving the homeless. *PBS.org*, January 28, 2015. Retrieved on April 7, 2016, from http://www.pbs.org/newshour/rundown/see-libraries-across-country-serving-homeless/.

Fuller, M., and A. Goffey (2012). *Evil media*. Cambridge, MA: MIT Press.

Fuller, W. (1964). *R. F. D.: The changing face of rural America*. Bloomington: Indiana University Press.

Fuller, W. (1972). *The American mail: Enlarger of the common life*. Chicago: University of Chicago Press.

Gabel, R. (1969). The early competitive era in telephone communication, 1893–1920. *Law and Contemporary Problem* 34(2): 340–359.

Gabriel. R. H. (1964). Rise of American industrial civilization. In *The history of American education through readings*, eds. C. H. Gross and C. C. Chandler, 212–221. Boston: D. C. Heath.

Garceau, O. (1949). *The public library in the political process: A report of the Public Library Inquiry*. New York: Columbia University Press.

Gardner, B. L. (2009). *American agriculture in the twentieth century: How it flourished and what it cost*. Cambridge, MA: Harvard University Press.

Garnet, R. W. (1985). *The telephone enterprise: The evolution of the Bell System's horizontal structure, 1876–1909*. Baltimore, MD: Johns Hopkins University Press.

Garrison, D. (1979). *Apostles of culture: The public librarian and American Society, 1876–1920*. New York: Free Press.

Geertz, C. (1973). *The interpretation of cultures: Selected essays*. New York: Basic Books.

Gehlen, A. (1980). *Man in the age of technology*. New York: Columbia University Press.

Gerdeman, D. (2018). Should retailers match their own prices online and in stores? *Forbes.com*, June 7. Retrieved on July 12, 2020, from https://www.forbes.com/sites/hbsworkingknowledge/2018/06/07/should-retailers-match-their-own-prices-online-and-in-stores/#4843d2fc9a3b.

German Law Archive (n.d.). Broadcasting law in Germany. Retrieved on June 13, 2018, from https://germanlawarchive.iuscomp.org/?p=386.

Gleick, J. (2012). *The information: A history, a theory, a flood*. New York: Vintage Books.

Good, H. G. (1956). *A history of American education*. New York: Macmillan.

Gordon, L. A., M. P. Loeb, W. Lucyshyn, and L. Zhou (2015). Increasing cybersecurity investments in private sector firms. *Journal of Cybersecurity* 1(1): 3–17.

Graham, S., and S. Marvin (2003). *Splintering urbanism: Networked infrastructures, technological mobilities and the urban condition.* London: Routledge.

Grand Forks Daily Herald (1900). Rural Free Delivery. *Grand Forks Daily Herald*, December 13, 4.

Grand Rapids Herald (1899). Benefit of rural mail: It has enhanced the value of farm property. *Grand Rapids Herald*, November 5, 1.

Grand Rapids Press (1916). U.S. demands good road for Rural Free Delivery. *Grand Rapids Press*, January 29, 9.

Gregory, J. N. (2005). *Southern diaspora: How the great migrations of black and white Southerners transformed America.* Chapel Hill: University of North Carolina Press.

Grossman, J. R. (1991). *Land of hope: Chicago, Black Southerners, and the Great Migration.* Chicago: University of Chicago Press.

Guback, T. H., and S. P. Hill (1972). The beginnings of Soviet broadcasting and the role of V. I. Lenin. *Journalism Monographs* 26: 1–43.

Gunderman, R., and D. C. Stevens (2015). How libraries became the front line of America's homelessness crisis. *Washington Post*, August 19. Retrieved on April 7, 2016, from https://www.washingtonpost.com/posteverything/wp/2015/08/19/how-libraries -became-the-front-line-of-americas-homelessness-crisis/?utm_term=.03b51566051b.

Guyot, K., and I. V. Sawhill (2020). Telecommuting will likely continue long after the pandemic. Brookings.edu, April 6. Retrieved on July 11, 2020, from https://www .brookings.edu/blog/up-front/2020/04/06/telecommuting-will-likely-continue-long -after-the-pandemic/.

Handlin, D. (1979). *The American home: Architecture and society, 1815–1915.* Boston: Little, Brown.

Hargittai, E. (2002). Second-level digital divide: Differences in people's online skills. *First Monday* 4(1). Retrieved on July 2, 2019, from https://firstmonday.org/ojs/index .php/fm/article/viewArticle/942/864Head.

Hargittai, E., and A. Shaw (2015). Mind the skills gap: The role of Internet know-how and gender in differentiated contributions to Wikipedia. *Information, Communication & Society* 18(4): 424–442.

Harlow, A. F. (1926). *Old towpaths: The American canal era.* New York: D. Appleton & Co.

Harper, D. (1992). Small N's and community case studies. In *What is a case? Exploring the foundations of social inquiry*, eds. C. C. Ragin and H. S. Becker, 139–158. New York: Cambridge University Press.

Harris, M. H. (1975). *The role of the public library in American life: A speculative essay.* Occasional paper. Urbana-Champaign: Graduate School of Library Science, University of Illinois.

Harris, M. H., and G. Spiegler (1974). Everett, Ticknor and the common man; the fear of societal instability as the motivation for the founding of the Boston Public Libraries. *Libri* 24(4): 249–276.

Hecht, B., and M. Stephens (2016). A tale of cities' urban biases in volunteered geographical information. In *Proceedings of the Eighth International Conference on Weblogs and Social Media*, 197–205. Retrieved on June 28, 2019, from https://www.aaai.org/ocs/index.php/ICWSM/ICWSM14/paper/viewPaper/8114.

Heidenry, M. (2011). My parents were home-schooling anarchists. *New York Times Magazine*, November 8. Retrieved on November 21, 2020, from https://www.nytimes.com/2011/11/13/magazine/my-parents-were-home-schooling-anarchists.html?pagewanted=all.

Hijiya, J. (1992). *Lee de Forest and the fatherhood of radio.* Cranbury, NJ: Associated University Press.

Hill, A. (1902). The public library and the people. *Library Journal* 27(1): 11–15.

Hillesheim, J. W., and G. D. Merrill (1971). *Theory and practice in the history of American education: A book of readings.* Pacific Palisades, CA: Goodyear.

Hochfelder, D. (2002). Constructing an industrial divide: Western Union, AT&T, and the federal government, 1876–1971. *Business History Review* 76(4): 705–732.

Hoffman, D. L., and T. P. Novak (1998). Bridging the racial divide on the Internet. *Science* 280(5362): 390–391.

Holland, J. D. (1938). Telephone service essential to progressive farm home. *Telephony*, February 19, 17–20.

Horwitz, R. B. (1989). *The irony of regulatory reform: The deregulation of American telecommunications.* New York: Oxford University Press.

Howard, B., and K. Han (2020). Millions of Africans lack access to clean water. This makes coronavirus a bigger threat. *Washington Post*, March 22. Retrieved on July 18, 2021, from https://www.washingtonpost.com/politics/2020/03/22/millions-africans-lack-access-clean-water-this-makes-coronavirus-bigger-threat/.

Hu, T-H. (2015). *A prehistory of the cloud.* Cambridge, MA: MIT Press.

Hughes, T. (1998). *Rescuing Prometheus: Four monumental projects that changed the modern world.* New York: Pantheon Books.

Hughes, T. P. (1983). *Networks of power: Electrification in Western society, 1880–1930.* Baltimore: Johns Hopkins University Press.

Hutsinpillar, C. A. (2012). The parcels post. In *The American postal network, 1792–1914, Vol. 4*, ed. R. R. John, 259–266. London: Pickering & Chatto.

IBISWorld (2019). Identity theft protection services industry in the US: Market research report. Retrieved on July 6, 2020, from https://www.ibisworld.com/united-states /market-research-reports/identity-theft-protection-services-industry.

Idaho Daily Statesman (1910). Big deficit is materially cut down: Annual report of postmaster general shows gratifying gain financially. *Idaho Daily Statesman*, December 12, 1.

Indianapolis Star (1909). Demands better roads or no Rural Free Delivery. *Indianapolis Star*, October 6, 1.

Innis, H. A. (1950). *Empire and communications*. Oxford, UK: Clarendon Press.

Innis, H. A. (1951). *The bias of communication*. Toronto: University of Toronto Press.

Institute of Museum and Library Services (2018). *Public libraries in the United States: Fiscal year 2015*. Washington, DC: Institute of Museum and Library Services.

Institute of Museum and Library Services (2019a). *Public libraries in the United States: Fiscal year 2016*. Washington, DC: Institute of Museum and Library Services.

Institute of Museum and Library Services (2019b). *Public libraries in the United States: Fiscal year 2017, Volume 1*. Washington, DC: Institute of Museum and Library Services.

Institute of Museum and Library Services (2020). *Public libraries in the United States: Fiscal year 2017, Supplementary Tables*. Washington, DC: Institute of Museum and Library Services.

Irving, L. (2001). Michael Powell's "Mercedes divide." *Washington Post*, June 30, A30.

Isaacs, J. (2008). *The costs of benefit delivery in the food stamp program: Lessons from a cross-program analysis*. Contractor and Cooperator Report No. 39. Washington, DC: United States Department of Agriculture.

Jackson, S. L. (1941). *America's struggle for free schools: Social tension and education in New England and New York, 1827–42*. Washington, DC: American Council on Public Affairs.

Jaffe, A. J., and W. Adams (1972). Two models of open enrollment. In *Universal higher education: Costs, benefits, options*, eds. L. Wilson and O. Mills, 223–251. Washington, DC: American Council on Education.

Jayakar, K. (2017). Universal broadband: Option, right or obligation? *Journal of Human Values* 24(1): 11–24.

Jayakar, K., and H. Sawhney (2004). Universal service: Beyond established practice to possibility space. *Telecommunications Policy* 28(3–4): 339–357.

Jellison, K. (1993). *Entitled to power: Farm women and technology, 1913–1963*. Chapel Hill: University of North Carolina Press.

Joeckel, C. B. (1939). *The government of the American public library*. Chicago: University of Chicago Press.

John, R. R. (1995). *Spreading the news: The American postal system from Franklin to Morse.* Cambridge, MA: Harvard University Press.

John, R. R. (2010). *Network nation: Inventing American telecommunications.* Cambridge, MA: Harvard University Press.

John, R. R. (2012). Introduction: Reform, 1861–1914. In *The American postal network, 1792–1914, Vol. 4,* ed. R. R. John, vii–xiv. London: Pickering & Chatto.

Johnson, I. L. (2016). Not at home on the range: Peer production and the urban/rural divide. In *CHI'16: Proceedings of the 2016 CHI Conference on Human Factors in Computing Systems,* 13–25. New York: ACM.

Jones, C. L. (1914). The parcel post in foreign countries. *Journal of Political Economy* 22(6): 509–525.

Kaestle, C. F. (1983). *Pillars of the republic: Common schools and American society, 1780–1860.* New York: Hill and Wang.

Kaestle, C. F., and M. A. Vinovskis (1980). *Education and social change in nineteenth-century Massachusetts.* New York: Cambridge University Press.

Kalamazoo Gazette (1908). Good roads in Kalamazoo: Rural Free Delivery has done much for Kalamazoo County. *Kalamazoo Gazette,* September 23, 3.

Kalisch, P. A. (1969). *The Enoch Pratt Free Library: A social history.* Metuchen, NJ: Scarecrow Press.

Karabel, J. (1972). Perspectives on open admissions. In *Universal higher education: Costs, benefits, options,* eds. L. Wilson and O. Mills, 265–286. Washington, DC: American Council on Education.

Katz, M. S. (1976). *A history of compulsory education laws.* Bloomington, IN: Phi Delta Kappa Educational Foundation.

Kelderman, E. (2014). Free community college? Tennessee proposal draws praise and concerns. *Chronicle of Higher Education,* February 6. Retrieved on July 12, 2018, from https://www.chronicle.com/article/Free-Community-College-/144553.

Kemper, W. H. (1908). *A twentieth century history of Delaware County, Indiana, Vol. 1.* Chicago: Lewis Publishing Company.

Kevane, M., and W. A. Sundstrom (2014). The development of public libraries in the United States, 1870–1930. *Information & Culture* 49(2): 117–144.

Kielbowicz, R. B. (1994a). Government goes into business: Parcel post in the nation's political economy, 1880–1915. *Studies in American Political Development* 8(1): 150–172.

Kielbowicz, R. B. (1994b). Rural ambivalence toward mass society: Evidence from the U.S. parcel post debates, 1900–1913. *Rural History* 5(1): 81–102.

Kireyev, P., V. Kumar, and E. Ofek (2017). Match your own price? Self-matching as a retailer's multichannel pricing strategy. *Marketing Science* 36(6): 813–1017.

Kirkpatrick, B. (2006). *Localism in American media, 1920–1934.* Doctoral dissertation, University of Wisconsin, Madison.

Kirkpatrick, B. (2011). Regulation before regulation: The local-national struggle for control of radio regulation in the 1920s. *Journal of Radio & Audio Media* 18(2): 248–262.

Klein, H. (2006). *Public access television: A radical critique.* Paper presented at the Tele-communications Policy Research Conference (TPRC), Washington, DC, September.

Kline, R. R. (2000). *Consumers in the country: Technology and social change in rural America.* Baltimore, MD: Johns Hopkins University Press.

Klinenberg, E. (2018). *Palaces for the people: How social infrastructure can help fight inequality, polarization, and the decline of civic life.* New York: Broadway Books.

Klotz, A. (2008). Case selection. In *Qualitative methods in international relations,* eds. A. Klotz and D. Prakash, 43–58. London: Palgrave Macmillan.

Knake, R. K. (2017). A cyberattack on the U.S. power grid (Contingency Planning Memorandum No. 31). New York: Center for Preventive Action, Council on Foreign Relations.

Knuth, R. (2003). *Libricide: The regime-sponsored destruction of books and libraries in the twentieth century.* Westport, CT: Praeger.

Kolowich, S. (2016). How a syllabus sparked a war between a professor and his college. *Chronicle of Higher Education,* April 8, A27–30.

Kotin, L., and W. F. Aikman (1980). *Legal foundations of compulsory student attendance.* Port Washington, NY: Kennikat Press.

Krasnow, E. G., and J. N. Goodman (1998). The "public interest" standard: The search for the Holy Grail. *Federal Communications Law Journal* 50(3): 606–636.

Krug, E. A. (1966). *Salient dates in American education, 1635–1964.* New York: Harper & Row.

Kurtz, C. F., and D. J. Snowden (2003). The new dynamics of strategy: Sense-making in a complex and complicated world. *IBM Systems Journal* 42(3): 462–483.

Landi, H. (2020). Half of physicians now using telehealth as COVID-19 changes practice operations. Retrieved on July 12, 2020, from https://www.fiercehealthcare.com /practices/half-physicians-now-using-telehealth-as-covid-changes-practice-operations.

Larned, J. N. (1894). Address of the president. In *Papers and Proceedings of the Sixteenth General Meeting of the American Library Association held at Lake Placid, NY, September 17–22,* 1–4. Chicago: American Library Association.

Lasch, C. (1991). *The true and only heaven: Progress and its critics.* New York: W. W. Norton.

Latzko-Toth, C., C. Bonneau, and M. Millette (2017). Small data, thick data: Thickening strategies for trace-based social media research. In *SAGE handbook of social media research methods,* eds. L. Sloan and A. Quan-Haase, 199–214. London: SAGE.

Lavin, D. E., R. D. Alba, and R. A. Silberstein (1981). *Right versus privilege: The open-admissions experiment at the City University of New York*. New York: Free Press.

Law, M. (1996). Celebrating 100 years of Rural Free Delivery. *New Jersey Postal History* 24(4): 100–123.

Leiner, B. M., G. C. Vinton, D. D. Clark, R. E. Kahn, L. Kleinrock, D. C. Lynch, et al. (1997). Brief history of the Internet. Retrieved on March 10, 2018, from https://www .Internetsociety.org/Internet/history-Internet/brief-history-Internet.

Lenhart, A., and J. B. Horrigan (2003). Re-visualizing the digital divide as a digital spectrum. *IT & Society* 1(5), 23–39.

Lentz, R. G. (2000). The e-volution of the digital divide in the US: A mayhem of competing metrics. *Info* 2(4): 355–377.

Lerner, F. (2009). *The story of libraries: From the invention of writing to the computer age*. 2nd ed. New York: Continuum.

Library Journal (1877). Editorial. *Library Journal* 1: 395–396.

Lieberman, J. L. (2017). *Power lines: Electricity in American life and letters, 1882–1952*. Cambridge, MA: MIT Press.

Lilienthal, D. E. (1939). Electricity: The people's business. *Annals of the American Academy of Political and Social Sciences* 201: 58–63.

Lipartito, K. (1989). *The Bell System and regional business: The telephone in the South, 1877–1920*. Baltimore, MD: Johns Hopkins University Press.

Loader, B. D., and L. Keeble (2004). *Challenging the digital divide? A literature review of community informatics initiatives*. York, UK: Joseph Roundtree Foundation.

Losh, S. C. (2003). Gender and educational digital chasms in computer and Internet access and use over time: 1983–2000. *IT & Society* 1(4): 73–86.

Lund, C. (2014). Of what is this a case? Analytical movements in qualitative social science research. *Human Organization* 73(3): 224–234.

Lurie, J. (2012). *William Howard Taft: The travails of a progressive conservative*. New York: Cambridge University Press.

MacDougall, R. D. (2004). *The people's telephone: The politics of telephony in the United States and Canada, 1876–1926*. Doctoral dissertation, Harvard University, Cambridge, MA.

Madsen, D. (1974). *Early national education: 1776–1830*. New York: Wiley.

Mann, H. (1964). An educator speaks on education. In *The history of American education through readings*, eds. C. H. Gross and C. C. Chandler, 94–101. Boston: D. C. Heath & Co.

Manning, R. (2015). Thousands to benefit from Oregon free community college bill. OPB.org (Oregon Public Broadcasting), July 17. Retrieved on July 12, 2018, from https://www.opb.org/news/article/governor-brown-signs-off-on-free-community-college/.

Marciano, C. (2020). How much does cyber/data breach insurance cost? Retrieved on July 7, 2020, from https://databreachinsurancequote.com/cyber-insurance/cyber-insurance-data-breach-insurance-premiums/.

Margolis, M., and D. R. Resnick (2000). *Politics as usual: The cyberspace "revolution."* London: SAGE.

Marvin, C. (1988). *When old technologies were new.* New York: Oxford University Press.

Mason, G. (1921). Putting the talk in Chautauqua. *The Outlook*, June 6, 418–420.

Mazzocato, L. (2020). The digital divide: Ethics and technology for everyone. The Digital Transformation People. Retrieved on July 4, 2021, from https://www.thedigitaltransformationpeople.com/channels/people-and-change/the-digital-divide-ethics-and-technology-for-everyone.

McCarthy, N. (2017). The most and least expensive countries for broadband. *Forbes.com*, November 22. Retrieved on April 25, 2022 from https://www.forbes.com/sites/niallmccarthy/2017/11/22/the-most-and-least-expensive-countries-for-broadband-infographic/?sh=693540ad23ef

McChesney, R. W. (1993). *Telecommunications, mass media, and democracy: The battle for the control of U.S. broadcasting, 1928–1935.* Oxford, UK: Oxford University Press.

McCook, K. de la Peña. (2001). *Poverty, democracy and public libraries.* School of Information Faculty Publications 111. Tampa: School of Information, University of South Florida. Retrieved on March 22, 2020, from http://scholarcommons.usf.edu/si_facpub/111.

McCrary, J. B. (1939). An engineer looks at REA. *Rural Electrification News*, September, 5.

McCraw, T. (1971). *TVA and the power fight, 1933–1939.* Philadelphia: J. B. Lippincott.

McNeely, I. F., and L. Wolverton (2008). *Reinventing knowledge: From Alexandria to the Internet.* New York: W. W. Norton.

Mechanic's Free Press. (1830). Untitled article. *Mechanic's Free Press*, January 23, 3.

Meinig, D. W. (2004). *The shaping of America: A geographical perspective on 500 years of history, Volume 4.* New Haven, CT: Yale University Press.

Mencken, H. L. (1955). Travail. In *The Vintage Mencken*, ed. A. Cooke, 182–185. New York: Vintage.

Merrefield, C. (2020). Rural broadband in the time of coronavirus. *Journalist's Resource*, March 30. Retrieved on July 11, 2020, from https://journalistsresource.org/studies/society/internet/rural-broadband-coronavirus/.

Messere, F. (2005). The Davis Amendment and the Federal Radio Act of 1927: Evaluating external pressures in policy making. In *Transmitting the past: Historical and cultural perspectives on broadcasting,* eds. J. E. Winn and S. L. Brinson, 34–68. Tuscaloosa: University of Alabama Press.

Messerli. J. (1967). The Columbian complex: The impulse to national consolidation. *History of Education Quarterly* 7(4): 417–431.

Meyer, A. E. (1965). *An educational history of the Western world.* New York: McGraw-Hill.

Michigan Farmer (1900). How to obtain Rural Free Delivery: Form of Petition. *Michigan Farmer,* February 3, 98.

Miles, S. R. (2012). The menace of a parcels post. In *The American postal network, 1792–1914, Vol. 4,* ed. R. R. John, 291–329. London: Pickering & Chatto.

Min, S-J. (2010). From the digital divide to the democratic divide: Internet skills, political interest, and the second-level digital divide in political Internet use. *Journal of Information Technology & Politics* 7(1): 22–35.

Mitchell, R. (2018). Andrew Carnegie built 1,700 public libraries. But some towns refused the steel baron's money. *Washington Post,* April 9. Retrieved on June 13, 2019, from https://www.washingtonpost.com/news/retropolis/wp/2018/04/09/andrew-carneg ie-built-1700-public-libraries-but-some-towns-refused-the-steel-barons-money/?utm _term=.613db3084744.

Mock, R. (2015). Ditch paper tax returns: The IRS could save billions by moving to mandatory e-filing. *U.S. News & World Report,* April 10. Retrieved on July 9, 2020, from https://www.usnews.com/opinion/economic-intelligence/2015/04/10/irs-could -save-billions-with-mandatory-e-filing.

Morning Herald (1902). Better boxes must be provided on rural routes or service will be withdrawn. *The Morning Herald,* December 22, 9.

Morning Oregonian (1903). Rural boxes condemned: Rural Free Delivery denied until appropriate boxes are erected. *Morning Oregonian,* February 25, 10.

Morozov, E. (2014). Facebook's gateway drug. *New York Times,* August 2. Retrieved on July 9, 2015, from http://www.nytimes.com/2014/08/03/opinion/sunday/evgeny -morozov-facebooks-gateway-drug.html?_r=0.

Morrow, J. B. (1909). Rural Free Delivery costs millions, while it saves hundreds of millions. *Plain Dealer,* December 5, 49.

Morrow, S. L. (2005). Quality and trustworthiness in qualitative research in counseling psychology. *Journal of Counseling Psychology* 52(2): 250–260.

Mosco, V. (2004). *The digital sublime: Myth, power, and cyberspace.* Cambridge, MA: MIT Press.

Mosher, W. E., and F. G. Crawford (1932). The economic importance of bringing current to the countrymen. *Public Utilities Fortnightly* 9(6): 330–336.

Mossberger, K., C. J. Tolbert, and R. S. McNeal (2008). *Digital citizenship: The Internet, society, and participation.* Cambridge, MA: MIT Press.

Mueller, M. (1989). The switchboard problem: Scale, signaling, and organization in manual telephone switching, 1877–1987. *Technology and Culture* 30(3): 534–560.

Mueller, M. (1993). Universal service in history: A reconstruction. *Telecommunications Policy* 17(5): 352–369.

Mueller, M. (1997). *Universal service: Competition, interconnection, and monopoly in the making of the American telephone system.* Cambridge, MA: MIT Press.

Mulgan, G. J. (1991). *Communication and control: Networks and the new economies of communication.* New York: Guilford.

Müller, M. (2016). Assemblages and actor-networks: Rethinking socio-material power, politics and space. *Geography Compass* 9(1): 27–41.

Mulrooney, J. B. (1937). Some aspects of rural telephone service. *Telephony*, January 9, 14–16.

Munday, L. A., and P. R. Rever (1971). Perspectives on open admission. In *Open admissions and equal access,* ed. P. R. Rever, 75–96. Iowa City, IA: American College Testing Program.

Munson, R. (1985). *The power makers: The inside story of America's biggest struggle to control tomorrow's electricity.* Emmaus, PA: Rodale Press.

Napoli, P. M. (2001). *Foundations of communications policy.* Cresskill, NJ: Hampton Press.

Nardini, R. F. (2001). A search for meaning: American library metaphors, 1876–1926. *Library Quarterly* 71(2): 111–140.

National Archives and Records Administration (2020). *Federal register,* Vol. 85, No. 113, June 11, 35567–35572. Washington, DC: United States Government Printing Office.

National Center for Education Statistics (2018). *Digest of education statistics,* Table 206.10. Retrieved on November 21, 2020, from https://nces.ed.gov/programs/digest /d18/tables/dt18_206.10.asp.

National Governors Association (1988). *A hearing of the subcommittee in telecommunications: Universal service, summary transcript.* Washington, DC: National Governors' Association.

National Telecommunications and Information Administration (NTIA) (1995). *Falling through the net: A survey of "have-nots" in rural and urban America.* Washington, DC: NTIA.

National Telecommunications and Information Administration (NTIA) (1999). *Falling through the net: Defining the digital divide.* Washington, DC: NTIA.

Nelson, S. (1982). Wealth classifications and equal protection: Quo vadimus? *Houston Law Review* 19(4): 713–731.

Newhagen, J. E., and E. P. Bucy (2004). Routes to media access. In *Media access: Social and psychological dimensions of new technology use,* eds. E. P. Bucy and J. E. Newhagen, 3–23. Mahwah, NJ: Lawrence Erlbaum.

New York Edison Company (1913). *Thirty years of New York, 1882–1912: Being a history of electrical development in Manhattan and the Bronx.* New York: Press of the New York Edison Company.

New York Times. (1898). Fences for telephone lines: New Jersey farmers to put the wires to practical use. *New York Times,* June 2, 4.

New York Times. (1899). Rural free mail delivery. *New York Times,* November 6, 3.

New York Times (1900a). Needed postal reforms. *New York Times,* December 10, 7.

New York Times (1900b). Rural Free Delivery. *New York Times,* December 12, 8.

New York Times (1900c). Rural Free Delivery: Popularity of the law shown in debate on deficiency bill. *New York Times,* January 19, 1.

New York Times (1901a). Blow to postal box trust. *New York Times,* July 17, 2.

New York Times (1901b). An improved "rural delivery." *New York Times,* September 29, 14.

New York Times (1901c). Postal department and boxmakers' trust. *New York Times,* January 13, 1.

New York Times (1901d). Rural weather forecasts: Postmen of free delivery routes to carry signal flags and to distribute printed predictions. *New York Times,* August 14, 6.

New York Times (1901e). Want free delivery stopped. *New York Times,* January 4, 3.

New York Times (1902a). Barbed wire phones: Montana ranchmen making a general use of the fences. *New York Times,* June 1, 18.

New York Times (1902b). Ex-postmaster general Bissell against government ownership. *The New York Times,* January 26, 2.

New York Times (1902c). Rural delivery aids merchants: Will be able to get more accurate mailing lists. *New York Times,* December 28, 10.

New York Times (1902d). Rural Free Delivery crowds out postmasters. *New York Times,* December 14, 27.

New York Times (1902e). Rural Free Delivery killing country trade. *New York Times,* February 9, 5.

New York Times (1902f). Traveling libraries: What women's clubs have done in the various states to foster these useful institutions. *New York Times,* January 19, 41.

New York Times (1904). Rural parcels post plan. *New York Times,* December 12, 6.

New York Times (1905). Fifty new daily papers: Rural Free Delivery responsible for the demand in Iowa. *New York Times,* July 26, 1.

New York Times (1907). The passing of rusticity. *New York Times,* February 3, 41.

New York Times (1908a). Mails wait on society. *New York Times,* March 5, 1.

New York Times (1908b). Rural Free Delivery grows. *New York Times*, December 1, 8.

New York Times (1909a). Delivering mail in our farthest north. *New York Times*, August 15, 7.

New York Times (1909b). Hitchcock plans to end postal deficit. *New York Times*, December 27, 16.

New York Times (1910a). Defends rural delivery. Congressman Sims says postmaster general doesn't give it full credit. *New York Times*, February 27, 2.

New York Times (1910b). Hitchcock replies to the magazines. *New York Times*, February 28, 3.

New York Times (1910c). Urges higher postal rates on magazines. *New York Times*, December 12, 4.

New York Times (1911a). Magazines against a postal subsidy. *New York Times*, March 6, 3.

New York Times (1911b). Postmaster general favors parcel post. *New York Times*, August 26, 8.

New York Times (1924a). Charge postal loss to rural delivery. *New York Times*, December 28, 3.

New York Times (1924b). Radio broadcasting started at KDKA four years ago today. *New York Times*, November 2, 6.

New York Times (1926). Farmers prefer market reports to music, radio survey reveals. *New York Times*, January 10, 8.

New York Times (1955). Old-timers mark 60 years of R.F.D. *New York Times*, December 11, 138.

Nickolai, K. A. (1984). The AT&T divestiture: For whom will the Bell toll? *William Mitchell Law Review* 10(3): 507–529.

Noam, E. (1992). *Telecommunications in Europe.* New York: Oxford University Press.

Nora, S., and A. Minc (1980). *The computerization of society: A report to the president of France.* Cambridge, MA: MIT Press.

Nord, D. (1986). The ironies of communication technology: Why predictions of the future so often go wrong. *The Cresset* 49: 15–20.

Nordyke, L. T. (1938). "Rag-time Annie" on the party telephone line. *Telephony,* October 15, 24–26.

Nye, D. E. (1990). *Electrifying America: Social meanings of a new technology, 1880–1940.* Cambridge, MA: MIT Press.

OCLC (2008). *From awareness to funding: A study of library support in America (a report to OCLC membership).* Dublin, OH: OCLC.

OCLC and American Library Association (ALA) (2018). *From awareness to funding: Voter perceptions and support of public libraries in 2018 (summary report)*. Dublin, OH: OCLC.

Office of Technology Assessment (1989). *High performance computing & networking for science* (background paper). Washington, DC: United States Government Printing Office.

Ono, H., and M. Zavodny (2002). Race, Internet usage, and e-commerce. *Review of Black Political Economy* 30(3): 7–22.

Onuegbu, C. I. (2006). *Travails of a widow*. Enugu, Nigeria: Cecta Nig.

Patton, M. Q. (2015). *Qualitative research & evaluation methods: Integrating theory and practice*. 4th ed. Thousand Oaks, CA: SAGE.

Ornstein, A. C. (1974). An overview of the disadvantaged, 1900–1970. In *Rethinking educational equality*, eds. A. Kopan and H. Walberg, 1–10. Berkeley, CA: McCutchan.

Peck, L. (1972). Rural Free Delivery. In *Essays on Sussex County and New Jersey postal history*, ed. L. Peck, 62–70. Secaucus, NJ: New Jersey Postal History Society.

Pérez-Peña, R. (2014). Tennessee governor urges 2 free years of community college and technical school. *New York Times*, February 4. Retrieved on July 12, 2018, from https://www.nytimes.com/2014/02/05/education/tennessee-governor-urges-2-free -years-of-community-college-and-technical-school.html.

Pew Research Center (2015). *Libraries at the crossroads*. Washington, DC: Pew Research Center. Retrieved on March 22, 2020, from https://www.pewresearch.org /internet/2015/09/15/who-uses-libraries-and-what-they-do-at-their-libraries/.

Pew Research Center (2016). *Libraries 2016*. Washington, DC: Pew Research Center. Retrieved on March 22, 2020, from https://www.pewresearch.org/internet/2016/09 /09/libraries-2016/.

Phillips, R. A. (1939). Rural problem in Iowa can be solved. *Telephony*, May 27, 24–27.

Plain Dealer (1904). Can no longer act as agent. *Plain Dealer*, May 24, 3.

Ponemon Institute and IBM Security (2020). *Cost of a data breach report 2019*. Traverse City, MI: Ponemon Institute.

Ponterotto, J. G. (2006). Brief note on the origins, evolution, and meaning of the qualitative research concept "thick description." *The Qualitative Report* 11(3): 538–549.

Pope, N. (n.d.a.). 100 years of parcels, packages, and packets, oh my!—Congressional opposition. Retrieved on August 5, 2017, from https://postalmuseum.si.edu/parcelpost 100/p2.html.

Pope, N. (n.d.b.). 100 years of parcels, packages, and packets, oh my!—parcel post introduction. Retrieved on August 5, 2017, from https://postalmuseum.si.edu/parcel post100/index.html.

Pope, N. (n.d.c.). 100 years of parcels, packages, and packets, oh my!—the service in use. Retrieved on August 5, 2017, from https://postalmuseum.si.edu/parcelpost100/p5.html.

Postmaster-General (1892). *Report of the postmaster-general of the United States; being part of the message and documents communicated to the two houses of Congress at the beginning of the second session of the fifty-second Congress.* Washington, DC: Government Printing Office.

Postmaster General (1897). *Report of the postmaster general: For the fiscal year ended June 30, 1897.* Post Office Departmental Annual Reports. Washington, DC: Government Printing Office.

Postmaster General (1914). *Report of the postmaster general: For the fiscal year ended June 30, 1913.* Post Office Departmental Annual Reports. Washington, DC: Government Printing Office.

Post Office Department (1910). *Summary of the department's reply to the Periodical Publishers Association of America regarding second-class mail.* Washington, DC: Government Printing Office.

Pound, A. (1926). *The telephone idea: Fifty years after.* New York: Greenberg.

Powell, F. (2018). These states offer tuition-free college programs. *U.S. News & World Report*, February 1. Retrieved on July 12, 2018, from https://www.usnews.com/education/best-colleges/paying-for-college/articles/2018-02-01/these-states-offer-tuition-free-college-programs.

Prentice, A. E. (1973). *The public library trustee: Image and performance on funding.* Metuchen, NJ: Scarecrow Press.

Prentice, A. E. (2011). *Public libraries in the 21st century.* Santa Barbara, CA: Libraries Unlimited.

Preston, P., and R. Flynn (2000). Rethinking universal service: Citizenship, consumption norms, and the telephone. *The Information Society* 16(2): 91–98.

ProPublica (2013). FAQ: What you need to know about the NSA's surveillance program. *ProPublica.org*, August 5. Retrieved on July 8, 2020, from https://www.propublica.org/article/nsa-data-collection-faq.

Public Agenda (2006). *Long overdue: A fresh look at public leadership attitudes about libraries in the 21st century.* New York: Public Agenda.

Public Broadcasting Service (PBS) (n.d.). The Hayloft Gang re-aired July 1, 2014. Retrieved on July 17, 2018 from: http://program.lunchbox.pbs.org/program/hayloft-gang/.

Puranik, M. (2019). What is the cost of a data breach? *Forbes*, December 2. Retrieved on July 7, 2020, from https://www.forbes.com/sites/forbestechcouncil/2019/12/02/what-is-the-cost-of-a-data-breach/#1db63ea229e7.

Radder, N. J. (1939). What plumbing means to rural schools. *Rural Electrification News*, October, 2.

Radio Broadcast (1928). What radio owes to chain broadcasting. *Radio Broadcast*, May, 68.

Ragin, C. C. (1992a). "Casing" and the process of social inquiry. In *What is a case? Exploring the foundations of social inquiry*, eds. C. C. Ragin and H. S. Becker, 217–226. New York: Cambridge University Press.

Ragin, C. C. (1992b). Introduction: Cases of "what is a case?" In *What is a case? Exploring the foundations of social inquiry*, eds. C. C. Ragin and H. S. Becker, 1–17. New York: Cambridge University Press.

Ragin, C. C., and H. S. Becker (eds.) (1992). *What is a case? Exploring the foundations of social inquiry*. New York: Cambridge University Press.

Rapp, L. (1996). Public service or universal service? *Telecommunications Policy* 20(6): 391–397.

Reed, H. W. (1935). Rural electrification. *Electrical World*, June 8, 58–60.

Reich, D. R. (1963). Accident and design: The reshaping of German broadcasting under military government. *Journal of Broadcasting* 7(3): 191–207.

Ritchie, J., and J. Lewis (eds.) (2003). *Qualitative research practice: A guide for social science students and researchers*. London: SAGE.

Ritchie, J., J. Lewis, and G. Elam (2003). Designing and selecting samples. In *Qualitative research practice: A guide for social science students and researchers*, eds. J. Ritchie and J. Lewis, 77–108. London: SAGE.

Robinson, J. P., P. DiMaggio, and E. Hargittai (2003). New social survey perspectives on the digital divide. *IT & Society* 1(5): 1–22.

Roosevelt, F. D. (1932). Address of Governor Franklin D. Roosevelt, Municipal Auditorium, Portland, Oregon, September 21. (Transcript by White House stenographer from shorthand notes of the speech; Franklin D. Roosevelt: "The Great Communicator," The Master Speeches Files, 1898, 1910–1945, File No. 518, Franklin D. Roosevelt Presidential Library and Museum). Retrieved on August 8, 2021, from http://www.fdrlibrary.marist.edu/_resources/images/msf/msf00530.

Roper, D. C. (1917). *The United States Post Office: Its past record, present condition, and potential relation to the new world era*. New York: Funk and Wagnalls.

Rosch, E. (2001). "If you depict a bird, give it space to fly": Eastern psychologists, the arts, and self-knowledge. *SubStance* 30(1): 236–253.

Rossman, J. E., H. S. Astin, A. W. Astin, and E. H. el-Khawas (1975). *Open admissions at City University of New York: An analysis of the first year*. Englewood Cliffs, NJ: Prentice-Hall.

Rudgard, O. (2018). Why are Silicon Valley execs banning their kids from using social media? *The Telegraph*, November 1. Retrieved on July 14, 2020, from https://www.msn.com/en-ie/money/technology/why-are-silicon-valley-execs-banning-their-kids-from-using-social-media/ar-BBPcNZP.

Rudolph, F. (1965). Introduction. In *Essays on education in the early republic*, ed. F. Rudolph, vii–xxv. Cambridge, MA: Harvard University Press.

Rudolph, R., and S. Ridley (1986). *Power struggle: The hundred-year war over electricity*. New York: Harper & Row.

Ruhlmann, E. (2014). A home to the homeless. *American Libraries Magazine*, November 14. Retrieved on April 7, 2016, from http://americanlibrariesmagazine.org/2014/11/24/a-home-to-the-homeless/.

Rural Electrification Administration (1938). *Annual report of Rural Electrification Administration—1937*. Washington, DC: United States Government Printing Office.

Rural Electrification Administration (1990). *A brief history of the Rural Electric and Telephone Programs*. REA Bulletin 800–801. Washington, DC: United States Department of Agriculture.

Rural Electrification News (1935a). Administrator Cooke discusses distribution costs with mechanical engineers. *Rural Electrification News*, December, 10–11.

Rural Electrification News (1935b). REA to finance farm wiring. *Rural Electrification News*, December, 2–3.

Rural Electrification News (1936a). Missouri farmers find novel uses for electricity. *Rural Electrification News*, November, 25.

Rural Electrification News (1936b). World power conference discussion shows interest in farm power. *Rural Electrification News*, October, 15–16, 28.

Rural Electrification News (1937). 1936 record in rural electrification expected to be broken in 1937. *Rural Electrification News*, January, 3.

Rural Electrification News (1939a). Cost of service cut by cyclometer registers. *Rural Electrification News*, November, 21.

Rural Electrification News (1939b). Washing machine wringer used for shelling of peas. *Rural Electrification News*, October, 22–23.

Rural Radio (1938). Here are the winners: "What radio means to my family" contest. *Rural Radio*, May, 2.

Rush, B. (1965). Thoughts upon the mode of education proper in a republic. In *Essays on education in the early republic*, ed. F. Rudolph, 9–23. Cambridge, MA: Harvard University Press.

Ryle, G. (1971). *Collected papers, Volume 2: Critical essays*. New York: Hutchinson.

Saettler, L. P. (1990). *The evolution of American educational technology*. Englewood, CO: Libraries Unlimited.

Samarajiva, R., and P. Shields (1990a). Integration, telecommunications, and development: Power in the paradigms. *Journal of Communication* 40(3): 84–105.

Samarajiva, R., and P. Shields (1990b). Value issues in telecommunications resource allocation in the Third World. In *Telecommunications, values, and the public interest*, ed. S. Lundstedt, 227–253. Norwood, NJ: Ablex.

Sawhney, H. (1993). Circumventing the center: The realities of creating a telecommunications infrastructure in the USA. *Telecommunications Policy* 17(7): 504–516.

Sawhney, H. (1994). Universal service: Prosaic motives and great ideals. *Journal of Broadcasting & Electronic Media* 38(4): 375–395.

Sawhney, H. (2003). Universal service expansion: Two perspectives. *The Information Society* 19(4): 327–332.

Sawhney, H. (2009). Innovations at the edges: The impact of mobile technologies on the character of the Internet. In *Mobile technologies: From telecommunications to media*, eds. L. Hjorth and G. Goggin, 105–117. New York: Routledge.

Sawhney, H., and K. Jayakar (1999). Universal service: Migration of metaphors. In *Making universal service policy: Enhancing the process through multidisciplinary evaluation*, eds. B. Cherry, A. Hammond, and S. Wildman, 15–37. Mahwah, NJ: Lawrence Erlbaum.

Sawhney, H., and K. Jayakar (2007). Universal access. *Annual Review of Information Science and Technology* 41: 159–221.

Sawhney, H., and S. Lee (2005). Arenas of innovation: Understanding new configurational potentialities of communication technologies. *Media, Culture & Society* 27(3): 391–414.

Sawhney, H., V. R. Suri, and H. Lee (2010). New technologies and the law: Precedents via metaphors. *European Journal of Legal Studies* 2(3): 38–54.

Scheerder, A., A. van Deursen, and J. van Dijk (2017). Determinants of Internet skills, uses and outcomes: A systematic review of the second- and third-level digital divide. *Telematics and Informatics* 34(8): 1607–1624.

Schivelbusch, W. (2014). *The railway journey: The industrialization of time and space*. Berkeley: University of California Press.

Schmitz, J., E. M. Rogers, K. Phillips, and D. Paschal (1995). The Public Electronic Network (PEN) and the homeless in Santa Monica. *Journal of Applied Communication Research* 23(1): 26–43.

Schradie, J. (2011). The digital production gap: The digital divide and Web 2.0 collide. *Poetics* 39(2): 145–168.

Schrag, P. (1971). Open admissions to what? In *Open admissions and equal access*, ed. P. R. Rever, 49–53. Iowa City, IA: American College Testing Program.

Schrage, M. (2000). *Serious play: How the world's best companies simulate to innovate.* Boston: Harvard Business School Press.

Scientific American (1900). A cheap telephone service for farmers. *Scientific American* March 31, 196.

Seely, B. E. (1986). Railroads, good roads, and motor vehicles: Managing technological change. *Railroad History* 155: 35–63.

Selwyn, N. (2003). Apart from technology: Understanding people's non-use of information and communication technologies in everyday life. *Technology in Society* 25(1): 99–116.

Semuels, A. (2015). Free tuition is not enough. *The Atlantic*, October 15. Retrieved on July 12, 2018, from https://www.theatlantic.com/business/archive/2015/10/free -tuition/410626/.

Servon, L. (2002). *Bridging the digital divide: Technology, community and public policy.* London: Blackwell.

Shelton, B. K. (1976). *Reformers in search of yesterday: Buffalo in the 1890s.* Albany: State University of New York Press.

Shenk, D., A. L. Shapiro, and S. Johnson (1998). Technorealism overview. Retrieved on July 17, 2020, from https://www.technorealism.org/.

Shirky, C. (2008). *Here comes everybody: The power of organizing without organization.* New York: Penguin.

Shmurak, S. (2016). How to start a neighborhood tool share. Retrieved on July 13, 2020 from https://learn.eartheasy.com/articles/how-to-start-a-neighborhood-tool-share/.

Simama, J. (2020). America's moral obligation for universal broadband. Retrieved on July 4, 2021, from https://www.governing.com/now/americas-moral-obligation-for -universal-broadband.html.

Simpson, I. (2014). US libraries become front line in fight against homelessness. *Reuters.com*, July 17. Retrieved on April 21, 2022, from https://www.reuters.com /article/us-usa-homelessness-libraries/u-s-libraries-become-front-line-in-fight-against -homelessness-idUSKBN0FM16V20140717.

Slattery, H. (1940). *Rural America lights up.* Washington, DC: National Home Library Foundation.

Smith, L. G. (1931). Lessons from urban service applied to rural distribution. *Electrical World*, May 2, 816–822.

Smith, R., and R. Barry (2019). America's electric grid has a vulnerable back door—and Russia walked through it. *Wall Street Journal*, January 10. Retrieved on July 7, 2020,

from https://robbarry.org/assets/pdfs/Americas%20Electric%20Grid%20Has%20a%20
Vulnerable%20Back%20Door-and%20Russia%20Walked%20Through%20It%20
-%20WSJ.pdf.

Smith, S. H. (1965). Remarks on education. In *Essays on education in the early republic*, ed. F. Rudolph, 167–223. Cambridge, MA: Harvard University Press.

Smithsonian National Postal Museum (n.d.). Rural mailboxes. Retrieved on August 17, 2017, from https://postalmuseum.si.edu/exhibits/current/customers-and-communities/reaching-rural-america/rural-mailboxes.html.

Smulyan, S. (1994). *Selling radio: The commercialization of American broadcasting, 1920–1934.* Washington, DC: Smithsonian Institution Press.

Socolow, M. J. (2001). *To network a nation: N.B.C., C.B.S. and the development of national network radio in the United States, 1925–1950.* Doctoral dissertation, Georgetown University, Washington, DC.

Stake, R. E. (1995). *The art of case study research.* Thousand Oaks, CA: SAGE.

Standage, T. (1999). *The Victorian Internet: The remarkable story of telegraph and the nineteenth century's on-line pioneers.* New York: Berkley Books.

St. Austell, F. (1928). Direct selling by radio: Is it a menace to the retail business structure? *Radio Broadcast*, May, 58–60.

Stauter, M. (1973). *The Rural Electrification Administration, 1935–1945: A New Deal case study.* Doctoral dissertation, Duke University, Durham, NC.

Stegman, M., J. Lobenhofer, and J. Quinterno (2003). *The state of electronic benefit transfer (EBT).* Working paper. Chapel Hill: Center for Community Capitalism, University of North Carolina, Chapel Hill.

Stephens, M. (2013). Gender and the GeoWeb: Divisions in the production of user-generated cartographic information. *GeoJournal* 78: 981–996.

Sterling, C. H., and J. M. Kittross (1990). *Stay tuned: A concise history of American broadcasting, 2nd ed.* Belmont, CA: Wadsworth.

Stern, P. A., and D. Townsend (2006). *New Models for Universal Access to Telecommunications Services in Latin America.* Bogotá: REGULATEL.

Stevens, T. (1964). An appeal for tax-supported schools. In *The history of American education through readings*, eds. C. H. Gross and C. C. Chandler, 111–119. Boston: D. C. Heath & Co.

Stevenson, S. (2009). Digital divide: A discursive move away from the real inequities. *The Information Society* 25(1): 1–22.

Streeter, T. (1996). *Selling the air: A critique of the policy of commercial broadcasting in the United States.* Chicago: University of Chicago Press.

Strong, J. (1963). *Our country.* Cambridge, MA: Harvard University Press.

Strover, S. (2001). Rural internet connectivity. *Telecommunications Policy* 25(5): 331–347.

Strover, S. (2003). The prospects for broadband deployment in rural America. *Government Information Quarterly* 20(2): 95–106.

Swanson, E. B. (2020). How information systems came to rule the world: Reflections on the information systems field. *The Information Society* 36(2): 109–123.

Tarr, J. A. (1984). The evolution of the urban infrastructure in the nineteenth and twentieth centuries. In *Perspectives on urban infrastructure*, ed. R. Hanson, 4–66. Washington, DC: National Academies Press.

Taylor, A. (2016). Before "fake news," there was Soviet "disinformation." *Washington Post*, November 26. Retrieved on July 7, 2020, from https://www.washingtonpost .com/news/worldviews/wp/2016/11/26/before-fake-news-there-was-soviet -disinformation/.

Taylor, B. P. (2010). *Horace Mann's troubling legacy*. Lawrence: University of Kansas Press.

Terranova, T. (2000). Free labor: Producing culture for the digital economy. *Social Text* 18(2): 33–58.

Thomas, G. (2011). A typology for the case study in social science following a review of definition, discourse, and structure. *Qualitative Inquiry* 17(6): 511–521.

Thomas, M. J. (1967). Preface. In *Presidential statements on education: Excerpts from inaugural and state of the union messages*, 1789–1967, ed. M. J. Thomas, 3–44. Pittsburgh, PA: University of Pittsburgh Press.

Thompson, E. P. (1967). Time, work-discipline, and industrial capitalism. *Past and Present* 38: 56–97.

Toogood, A. (1978). West Germany: Federal structure, political influence. *Journal of Communication* 28(3): 83–89.

Trucano, M. (2015). Universal service funds & connecting schools to the internet around the world. World Bank Blogs (EduTech), February 26, 2015. Retrieved on July 16, 2021, from https://blogs.worldbank.org/edutech/universal-service-funds-con necting-schools-internet-around-world.

Universal Service Administrative Company (2020). *2019 annual report*. Washington, DC: Universal Service Administrative Company.

Universal Service Administrative Company (2021). *2020 annual report*. Washington, DC: Universal Service Administrative Company.

Urban, C. E. (1920a). The radio amateur. *Pittsburgh Gazette Times*, November 7, 3.

Urban, C. E. (1920b). The radio amateur: A department for wireless news. *The Pittsburgh Gazette Times*, October 24, 4.

Urban, C. E. (1920c). The radio amateur: A department for wireless news. *The Pittsburgh Gazette Times*, October 31, 10.

US Census Bureau (2012). The Great Migration, 1910 to 1970. Retrieved on July 31, 200, from https://www.census.gov/dataviz/visualizations/020/.

USPS Historian (2005). John Wanamaker. Retrieved on January 4, 2013, from http://about.usps.com/who-we-are/postal-history/pmg-wanamaker.pdf.

Vaidhyanathan, S. (2009). *The anarchist in the library*. New York: Basic Books.

Vaillant, D. (2002a). Sounds of whiteness: Local radio, racial formation, and public culture in Chicago, 1921–1935. *American Quarterly* 54(1): 25–66.

Vaillant, D. (2002b). "Your voice came in last night . . . but I thought it sounded a little scared": Rural radio listening and "talking back" during the Progressive Era in Wisconsin, 1920–1932. In *Radio reader: Essays in the cultural history of radio*, eds. M. Hilmes and J. Loviglio, 63–88. New York: Routledge.

Vaillant, D. (2004). Bare-knuckled broadcasting: Enlisting manly respectability and racial paternalism in the battle against chain stores, chain stations, and the Federal Radio Commission on Louisiana's KWKH, 1924–33. *Radio Journal* 1(3): 193–211.

Vaillant, D. (2017). *Across the waves: How the United States and France shaped the international age of radio*. Urbana: University of Illinois Press.

van Alstyne, M. (1997). The state of network organization: A survey in three frameworks. *Journal of Organizational Computing and Electronic Commerce* 7 (2–3): 83–151.

vanden Heuvel, K. (2020). America's digital divide is an emergency. *Washington Post*, June 23. Retrieved on July 4, 2021, from https://www.washingtonpost.com/opinions/2020/06/23/americas-digital-divide-is-an-emergency/.

Vanderbilt, T. (2020). Fortnite: Dad invasion. *Wired*, November, 28–33.

van der Mensbrugghe, F. (2003). *The codification of universal service: Converging answers to different tradition*. Doctoral dissertation, Université Catholique de Louvain, Louvain-la-Neuve, Belgium.

van Dijk, J. (2005). *The deepening divide: Inequality in the information society*. Thousand Oaks, CA: SAGE.

Wadlin, H. G. (1911). *The public library of the city of Boston: A history*. Boston: Boston Public Library.

Walker, I. (2019). Cybercriminals have your business in their crosshairs and your employees are in cahoots with them. *Forbes.com*, January 31. Retrieved on July 7, 2020, from https://www.forbes.com/sites/ivywalker/2019/01/31/cybercriminals-have-your-business-their-crosshairs-and-your-employees-are-in-cahoots-with-them/#d45 26931953d.

Washington Post (1910). Mr. Hitchcock discusses the cost of carrying magazines. *The Washington Post*, February 28, 1.

Washington Post (2001). Editorial: Michael Powell's "digital divide." *The Washington Post*, June 30, A30.

Whitehurst v. Grimes, 21 F.2d 787 (E.D. Ky. 192). Retrieved on June 10, 2018, from https://law.justia.com/cases/federal/district-courts/F2/21/787/1509996/.

Webb, L. D. (2006). *The history of American education: A great American experiment.* Columbus, OH: Peerson.

Weeks, P., and J. B. Gidney (1981). *Subjugation and dishonor: A brief history of the travail of native Americans.* Huntington, NY: Krieger.

Weisskopf, W. A. (1975). The dialectics of equality. In *The "inequality" controversy: Schooling and distributive justice,* eds. D. M. Levine and M. J. Bane, 214–227. New York: Basic Books.

Wellard, J. A. (1937). *Book selection.* London: Grafton.

White House (2021). Fact sheet: The American jobs plan. Retrieved on July 5, 2021, from https://www.whitehouse.gov/briefing-room/statements-releases/2021/03/31/fact -sheet-the-american-jobs-plan/.

Wickard, C. R. (1950). The REA telephone program. *Telephony,* October 21, 45–46, 111–112.

Wiegand, W. A. (2015). *Part of our lives: A people's history of the American public library.* New York: Oxford University Press.

Wieviorka, M. (1992). Case studies: History or sociology? In *What is a case? Exploring the foundations of social inquiry,* eds. C. C. Ragin and H. S. Becker, 159–72. New York: Cambridge University Press.

Wik, R. M. (1988). The USDA and the development of radio in rural America. *Agricultural History* 62(2): 177–188.

Willingham, W. W. (1970). *Free-access higher education.* New York: College Entrance Examination Board.

Winsor, J. (1876). Free libraries and readers. *Library Journal* 1: 63–67.

Wisconsin v. Yoder. 1972, 406 U.S. 205.

Wise, A. E. (1983). Educational adequacy: A concept in search of meaning. *Journal of Educational Finance* 8(3): 300–315.

Xenos, M., and P. Moy (2007). Direct and differential effects of the Internet on political and civic engagement. *Journal of Communication* 57(4): 704–718.

Xie, B. (2003). Older adults, computers, and the Internet: Future directions. *Gerontechnology* 2(4): 289–305.

Zetter, K. (2016). Everything we know about how the FBI hacks people. *Wired.com,* May 15. Retrieved on July 9, 2020, from https://www.wired.com/2016/05/history -fbis-hacking/.

Zinder, H. (1936). Municipal utilities urged to sell power to cooperatives on near-cost basis. *Rural Electrification News,* December, 14–16.

Zittrain, J. (2008). *The future of the internet and how to stop it?* New Haven, CT: Yale University Press.

Zuboff, S. (1988). *In the age of the smart machine.* New York: Basic Books.

Zuckerberg, M. (2012). The hacker way: Mark Zuckerberg's letter to investors. *CNN .com*, February 1. Retrieved on July 9, 2015, from http://money.cnn.com/2012/02 /01/technology/zuckerberg_ipo_letter/.

INDEX